Pilot Society and the Energy Transition

"This important book interrogates scholarship on sustainability transitions with insights from Science and Technology studies (STS) by focusing on pilot projects. The analysis is both critical towards dominant paradigms, and refreshingly constructive in the sense that it describes new ways of thinking about innovation practice and technology policy. The book urges us to look beyond technological solutionism, to examine how the energy transition also requires experimentation and even transformation in social domains. Pilot Society is a valuable contribution to discussions about how to make energy transitions just, fair and more humane, and it will be an important resource for students and scholars alike. Highly recommended!"
—Benjamin K. Sovacool, *University of Sussex and Aarhus University*

"The book provides a very useful review and knowledge synthesis of recent research and sets the work of the authors and their colleagues in a broader context. It is well-written, enjoyable to read and takes scholarship forwards in two ways: by presenting a clearly articulated perspective on pilots/demonstrations; and by providing a thorough review of literature on pilot projects from a broad perspective of how they are funded, imagined and enacted."
—Patrick Devine-Wright, *University of Exeter*

Marianne Ryghaug • Tomas Moe Skjølsvold

Pilot Society and the Energy Transition

The co-shaping of innovation, participation and politics

Marianne Ryghaug
Department of Interdisciplinary
Studies of Culture
Norwegian University of Science and
Technology (NTNU)
Trondheim, Norway

Tomas Moe Skjølsvold
Department of Interdisciplinary
Studies of Culture
Norwegian University of Science and
Technology (NTNU)
Trondheim, Norway

ISBN 978-3-030-61183-5   ISBN 978-3-030-61184-2   (eBook)
https://doi.org/10.1007/978-3-030-61184-2

© The Editor(s) (if applicable) and The Author(s) 2021, corrected publication 2021. This book is an open access publication.
Open Access This book is licensed under the terms of the Creative Commons Attribution 4.0 International License (http://creativecommons.org/licenses/by/4.0/), which permits use, sharing, adaptation, distribution and reproduction in any medium or format, as long as you give appropriate credit to the original author(s) and the source, provide a link to the Creative Commons licence and indicate if changes were made.
The images or other third party material in this book are included in the book's Creative Commons licence, unless indicated otherwise in a credit line to the material. If material is not included in the book's Creative Commons licence and your intended use is not permitted by statutory regulation or exceeds the permitted use, you will need to obtain permission directly from the copyright holder.
The use of general descriptive names, registered names, trademarks, service marks, etc. in this publication does not imply, even in the absence of a specific statement, that such names are exempt from the relevant protective laws and regulations and therefore free for general use.
The publisher, the authors and the editors are safe to assume that the advice and information in this book are believed to be true and accurate at the date of publication. Neither the publisher nor the authors or the editors give a warranty, expressed or implied, with respect to the material contained herein or for any errors or omissions that may have been made. The publisher remains neutral with regard to jurisdictional claims in published maps and institutional affiliations.

Cover pattern © Melisa Hasan

This Palgrave Pivot imprint is published by the registered company Springer Nature Switzerland AG.
The registered company address is: Gewerbestrasse 11, 6330 Cham, Switzerland

The original version of this book was revised. The book was inadvertently published with an incorrect surname to Tomas Moe Skjølsvold of captioned title as "Moe Skjølsvold" whereas it should be "Skjølsvold". The author's name has been updated in the book. The correction to this book can be found at https://doi.org/10.1007/978-3-030-61184-2_5

Contents

1 Transforming Society Through Pilot and Demonstration Projects — 1

2 The Co-production of Pilot Projects and Society — 23

3 Democratic and Participatory Pilot Projects? — 63

4 Catering for Socio-technical Transformations: Rethinking Technology Policy for Inclusive Transformation — 93

Correction to: Pilot Society and the Energy Transition — C1

References — 113

Index — 127

CHAPTER 1

Transforming Society Through Pilot and Demonstration Projects

Abstract This chapter introduces pilot and demonstration projects as a key mode of innovation within contemporary energy and mobility transitions. It argues that such projects are important political sites for the production of future socio-technical order. The politics of such projects are contested: on the one hand, they have been argued to remove political agency from deliberative fora in favour of private decisions, on the other hand they have been argued to constitute new democratic opportunities. This chapter situates a discussion on these issues within Science and Technology Studies (STS). The chapter further discusses the relationship between STS and some of the currently dominating approaches to sustainability transitions and argues how STS can bring new insights to the study of energy transitions and societal change. The chapter also provides basic insights into some key social and technical aspects of current energy and mobility transitions.

Keywords Pilot projects • Energy transitions • Sustainability transitions • STS • Participation • Politics

INTRODUCTION

In 1882, Thomas Edison unveiled a spectacular public display by electrically illuminating the offices of Drexel, Morgan and Company in the financial district of New York. Powered by the Pearl Street station in

Manhattan, this represented a challenge to what has been described as a "formidable rival", a gas industry and infrastructure that was not only economically and technically dominant, but an integral aspect of how the city functioned. In the words of Hargadon and Douglas, "*gas was inextricably woven into the city's physical and institutional environments*" (2001, p. 484). Nevertheless, incandescent light bulbs replaced gas in no more than 15 years, in what truly was a transition in the way New York was illuminated.

Many scholars have since noted that this success cannot be attributed to any single invention, and especially not the one made by Edison. Electric light bulbs had been successfully displayed decades before the 1882 events, and Edison was neither the inventor of the generator, nor the distribution system. The novelty in what Edison and his company did in New York could primarily be found in the systemic traits of their efforts, which combined power generation distributed through an electricity grid and in turn used by a small set of real-life users (see e.g. Hughes 1993; Hargadon and Douglas 2001; Geels 2010a). Hence, Edison illustrated real-world application for a new type of socio-technical configuration which has later been described as "*providing the model for subsequent development of the technology*" (Hargadon and Douglas 2001, p. 482).

Presently, we are faced with a situation where the need to transition away from large technical systems based on fossil fuels has become evident. In practice this means that a shift is needed away from the very coal power stations introduced by Edison, as well as the many technologies that provide heating, cooling, light, digital images, storage, industry, transport and all the other services we take for granted in late modernity that are enabled by the burning of fossil fuels. These challenges by far exceed those of replacing the gas system for illuminating New York in the 1880s and 1890s. The International Panel on Climate Change (Rogelj et al. 2018) highlights that keeping within the boundaries of 1.5 or 2.0-degree global warming will require a rapid transition of both supply and demand aspects of global energy systems by 2050. In practice, this means working to transform both energy production and energy use—and that this must happen quickly.

The grand challenges facing the energy- and sustainability transitions that need to happen are the core interest of this book. While the IPCC mainly operates macroscopically to illustrate how energy transitions need to pan out globally, our ambition is to zoom in, to take stock of and observe how transitions are enacted and how they unfold at various spatial scales, with different types of actor constellations, technologies and logics involved. As we zoom in, we also probe the work of transforming energy and transport systems as processes with wider implications than they would have had if the task was simply to replace light bulbs, generators, engines or energy carriers.

If Edison's trials had been conducted today, it would have been part of a broad movement, where social and technical configurations are actively "beta tested" in a limited way before being introduced to society at large (Marres 2020). Edison and his team conducted a small-scale trial under realistic conditions, which over the coming decades would be scaled up to provide blanket electricity coverage across nations and continents. Today, innovators, policy makers and research funders are actively pursuing and attempting to re-create similar dynamics within fields such as smart energy technology, renewable energy and electro mobility, through the establishment of pilot projects, test beds or demonstration projects. What such efforts entail and the effects of such projects are the key interests of this book.

Hargadon and Douglas (2001) noted that the effects of Edison's work in New York were not primarily technological, but institutional. Contemporary energy and transport projects are often made with the intention of testing how technology works in practice. In this book, however, we will make the point that such projects are always both social and technical: they do not only re-configure technological systems but also institutions, practices, everyday-life and politics. As the logics of piloting and pilot projects become more prevalent, such projects come to shape societies in new ways. Hence, the title of this book: *Pilot Society and the Energy Transition*, indicates that pilot projects have become one of the keyways through which societies are made and re-made. Such a perspective elevates the importance of innovation as an activity with far reaching consequences and opens for at least two types of questions. On the one hand, is the instrumental question of how such innovation can succeed. On the other hand, are questions of how such activities can be conducted in a fair, just and democratic way, promoting outcomes that not only reduce climate emissions but also produce just future societies.

Innovation and Politics Through Pilot and Demonstration Projects

Our key focus in this book is a specific form of innovation that has risen in prominence as a way of responding to climate and sustainability challenges (Hughes et al. 2018; Castan Broto and Bulkeley 2013). As noted, this type of innovation is enacted through projects that might be referred to as pilot projects, demonstration projects, experiments or test beds. Given the diversity of such activities it is difficult to accurately quantify how prevalent they are, but Castan Broto and Bulkeley (2013) surveyed 627 projects of this type across 100 cities. In the following years, research and innovation funders such as the Horizon 2020 have increasingly emphasized the importance of this mode of working, which means there is no reason to believe that the trend is fading (see e.g. European Commission 2020).

Throughout this book, we use terms such as pilot projects, demonstrations and experiments as synonyms. Further, our interpretation of what constitutes a pilot is broad. For us, this term includes relatively small projects, as well as larger, targeted sets of projects and policies that set out to explicitly create new socio-technical realities within a demarcated site. An example of the latter is the Norwegian effort to become a pioneering society for electro mobility. This is not a pilot project in the traditional sense, but it is a concerted push within a geographically and socially limited area to produce conditions that materialize visions held by policy makers and stakeholders about how a future dominated by electromobility might look. In this way, the country is also frequently discussed as a "laboratory" for transport electrification. Our pragmatic use of these words echoes the diverse ways that they are used amongst practitioners in the fields we study. Further, it signals that our interest does not lie in establishing a fine-grained typology of different activities, but rather more broadly, to explore a mode of innovation, which pits a set of ideas, technologies and principles of organizing innovative work against relatively realistic conditions and in what is often a public setting.

Coming from a background in Science and Technology Studies, the sorts of projects we discuss in this book can also be described as hybrids. Whereas scientific experiments have typically been conducted to learn about the character of the natural world, scientific demonstrations have been set-up to reveal such characteristics to a public audience (Latour 1983; Collins 1988). The sorts of projects discussed in this book as pilot projects, experiments or demo projects often do both. First, they intend to demonstrate to funders, commercial actors, researchers, governments and lay people, that some socio-technical configurations can work in real life and that they have desirable traits. Second, they often also seek to learn about how these socio-technical configurations interact with other elements in the world: how the configuration works in practice, how new technologies are used, what the consequences of new business models are, how the configuration interacts with different infrastructural systems and so on. Hence, pilot and demonstration projects often actively seek to produce a new reality, while at the same time studying the unfolding of this reality.

As hinted at, this means that the goals of such projects can be diverse. The goals are, however, often framed in terms of achieving quite specific technological goals (Marres 2018). Such technological goals can include testing and gaining practical experiences with the way new technologies work in (near to) real life conditions or investigating the complexities that might arise when several technologies are intended to work together. Other goals can be understood as to greater extent relate to social aspects of technology. Examples can include exploring how different actors understand or use technology or understanding why technologies are rejected. Projects can, for example, be rigged specifically for the purposes of generating social learning within an organization or amongst different actor groups, or to demonstrate or challenge what innovators consider to be flaws in current legislative and regulatory frameworks. At other times, such projects target the public, seeking to understand if emerging technologies, market structures and organizational forms are likely to be accepted, supported or even rejected.

We consider the development of pilot and demonstration projects to be a key strategy for enacting sustainability transitions in contemporary Europe and beyond, but we also believe that there is an untapped potential in doing such projects differently than they typically are today. While the paragraph above describes relatively conservative forms of innovation rooted in ideas of transforming society through implementing new technologies, there are also interesting examples of approaches that starts from a focus on social aspects. How can, for example, new energy technologies and new design practices be mobilized to produce new forms of communities (Martiskainen et al. 2018; Wilkie and Michael 2018)? In the following we will explore both how and why pilot projects are made, and consequences of carrying out these kinds of innovation projects once they are established. This means that we focus on the types of resources that are mobilized to make and shape such projects. Examples include European, national and local policies, a range of local issues, technologies and competence, as well as a diversity of actors.

Our focus on what demonstration and pilot projects *do* leads us to an interest in how they become part of broader societal transition processes. Such projects are seldom conducted without explicit ambitions of subsequent up-scaling, or of some form of transformative ambition that goes beyond the project as such (see e.g. Frantzeskaki et al. 2017; Naber et al. 2017; Ryghaug et al. 2019). If demonstration and pilot projects both formulate and materialize potential sustainable futures and succeed in transferring elements of such futures beyond their own immediate site and situation, they are important political entities that work to re-shape how key elements of contemporary and future societies are constituted. If societies indeed become pilot societies, such projects should be of major interest for social scientists, as sites that explicitly illustrate the constant making and re-making of society, hinting at potential directions and consequences before they are stabilized at a large scale.

Since we understand pilot and demonstration projects as political entities that are important for shaping the future not only for individual technologies but also, more broadly, for the societies that these technologies become part of, we are keen to explore the politics of such projects. An important aspect of this is the possibility that such projects might open for new modes of public participation in energy transition activities. On the one hand, we follow this question by enquiring into and against the backdrop of a quite common analysis that sees the implementation e.g. of smart energy technologies as a form of de-humanizing post-politics, or even anti-politics (Sadowski and Levenda 2020). Such analysis highlights that rapid technological change tends to result in the privatization of increasing aspects of societal decision making at the expense of traditional political institutions (e.g. Rosa 2013), and consequently that many technologies associated e.g. with the idea of smart energy or smart grids, limit the enactment of human agency in energy systems (Sadowski and Levanda 2020). On the other hand, we look into the potential for new modes of public participation in energy transition activities building on and being inspired by a body of literature that highlights the role of technologies and material objects in constituting issues and publics, and through this also enabling new forms of participation and new modes of democratic practice (e.g. Marres 2016).

When new forms of participation emerge, we are interested in understanding how this participation comes about. As we inquire into this issue, we are inspired by literature that highlights how participation emerges as a relational phenomenon within wide ecologies of actors (Chilvers and Kearnes 2015; Chilvers et al. 2018). Such a perspective points to that pilot and demonstration projects are not only sites where citizens, organizations, companies and researchers can opt-in or opt-out of participation in organized transition-oriented activities. They are sites where participation is formatted or orchestrated through the work of other actors and through the mobilization of ideas about human agency, technologies, research and innovation practices and policies (Skjølsvold et al. 2018). We are particularly interested in how the participation of citizens is orchestrated in such projects. Conventionally, such projects have tended to cast participation of citizens in the form of acting as consumers, attempting to instigate individual behaviour change, or producing acceptance for new technologies (Chilvers et al. 2018).

In some instances, however, other and more material, political and issue-oriented forms of participation emerge (Throndsen and Ryghaug 2015; Martiskainen et al. 2018). An example of this can be found in projects that enable the enactment of what we call energy citizenship (Ryghaug et al. 2018). In such instances, the materiality of projects anchored, for example in technologies like solar panels and electric vehicles enables new political virtues on behalf of citizens (see also Szulecki 2018). Examples of such virtues include the formation of *awareness*, the formation of *new knowledge* and literacy as well as new modes of *action* and practices. These elements can be directed towards the enactment of political projects such as advancing energy transitions, mitigating climate change or promoting equity. In sum, our observations suggest that pilot projects *can* play an important role in enabling new and democratic forms of transition, but this is far from any predetermined outcome.

What we sketch out above indicates that our interest in pilot and demonstration projects is operationalized through a socio-technical understanding of the dynamics of innovation, politics and participation towards sustainability transitions. Our account here is rooted in concepts and ideas primarily from Science and Technology Studies (STS), but in the discussions that follow in this book, we also borrow insights from other strands of social scientific scholarship on energy and sustainability transitions. In what follows, we will first discuss the theoretical underpinnings of our work, before proceeding to highlight some of the key (socio-technical) traits of contemporary energy transitions relevant to discussions raised in this book.

Sustainability Transitions: A Socio-technical Backdrop

The social sciences have always been concerned with understanding the relationship between social change and technological change. Johan Schot and Laur Kanger (2018) have argued that the last 250 years of industrialization and modernization can be described as a deep socio-technical transition, where the outcomes have been "increased labour productivity, mechanization, reliance on fossil fuels, resource-intensity, energy-intensity, and reliance on global value chains" (ibid., p. 1045). In light of such an interpretation of modernity we might read many of the classical social scientists such as Max Weber, Emile Durkheim, Georg Simmel and Karl Marx as primarily analysing social consequences of a long-term socio-technical transition (see also Rosa 2013 for a related argument).

Over the last decades, an explicit focus on transitions has emerged as a social scientific way of engaging with the challenges of climate change and sustainability. Scholars have addressed such transitions from different perspectives. In this book we are particularly relating to those perspectives that explore the social and technical aspects of such transitions in tandem; in other words, those perspectives rooted in what we can call a socio-technical understanding. Such perspectives have, on the one hand, been aimed at understanding the dynamics of transitions that have already unfolded in the past, while on the other hand, cultivating a normative and interventionist agenda aimed at understanding how to instigate contemporary and future transitions. In the following sections, we will first point to some of the dominant modes of such socio-technical analysis, before briefly outlining our own position.

From Multi-level Perspectives to Symmetrical Understandings of the Social and Technical Processes of Sustainability Transitions

In current academic discussions, the multi-level perspective (MLP) stands out as a particularly prominent framework for analysing sustainability transitions, with recent contributions in high profile journals like Science flagging ambitions far beyond any disciplinary boundaries (Geels et al. 2017). Building primarily on historical accounts (Geels 2002, 2005), the MLP makes a three-level conceptual distinction between niches, sociotechnical regimes and a landscape. The regime is arguably the key analytical concept of the MLP, as it represents a stable meso-level structure that contains the dominant "*products and technologies, stocks of knowledge, user practices, expectations, norms, regulations etc*" (Markard and Truffer 2008). Inspired by institutional theory, Geels (2004) posits that socio-technical regimes can be understood as a form of deep structure or grammar, or as a set of internally consistent rules that in evolutionary terms means that they are "Selection environment[s] for technological development in a certain field or sector, thus exerting a significant barrier for radical innovations to diffuse" (Markard and Truffer 2008).

Innovations can and do occur within socio-technical regimes. However, the literature highlights that such innovations tend to be incremental (e.g. Geels and Schot 2007), and hence insufficient in meeting current climate and sustainability challenges (Schot and Kanger 2018). Thus, through the lens of the MLP, sustainability transitions entail changing the character of existing regimes, or creating new regimes, mainly by way of creating radical breakthroughs of new niche technologies.

Analysed from a multi-level perspective, transitions emerge through interaction between the niche, regime and landscape level. Niches are the micro level and this is where scholars in this tradition typically identify radical innovations. For our purpose, niches are particularly interesting, since they have been highlighted as fertile soil for experimentation (e.g. Coenen et al. 2010). As part of this, pilot projects have been put forward as common elements in creating niche spaces (Raven et al. 2016). Niches tend to be organized as "protective spaces", which means that they serve to shield, nurture or empower new socio-technical configurations in order to strengthen their chances against the selection environment of established regimes (Smith and Raven 2012). Niches, then, tend to be described as smaller than regimes, and with rules that are less stable than those in regimes.

The macro landscape is largely seen as exogenous to the system. It is the "technical, physical and material backdrop that sustains society" (Geels and Schot 2007, p. 403). Change at this level is very slow, except for sudden shocks, such as wars, economic crises or pandemics. Despite this, landscapes change over time, and through this exert pressure on regimes.

In sum, through the lens of the MLP one can say that transitions emerge through:

> [...] External 'landscape' pressures (eg, climate change or cultural shifts) exerting pressure upon incumbent regimes (eg, the fossil-fuel based energy system) to open up 'windows of opportunity that might be filled by novel, radical innovations developed in 'niche' spaces (eg, renewable energy technologies). (Hargreaves et al. 2013, p. 403)

The journey of a new technology within such a scheme is often depicted as an s-curve, which tends to be described through four phases of transition (Rotmans et al. 2001). First, is a pre-development phase in which the status quo of socio-technical systems does not visibly change much. Second, is a take-off phase in which the state of the system begins to shift. Third, is an acceleration or breakthrough phase, where an accumulation of socio-cultural, economic, ecological and institutional changes react to each other, resulting in collective learning, diffusion and embedding of new technologies. Finally, a fourth phase, stabilization, is reached, and the speed of social change decreases. The journey is complete, and the niche technology has become part of an existing regime and thereby changes it, or a new regime is established. Studies using a multi-level perspective have tended to focus on the early phases of transitions, often in the form of experiments and pilot and demonstration projects. In this book we operate with a broader conceptualization, which renders experimentation visible also in later phases, and amongst actors who are not traditional niche actors.

As noted, the authors of this book come to the study of sustainability transitions from the perspective of Science and Technology studies (STS). The MLP, as discussed above, arguably represents a sort of synthesis of certain theoretical traditions within STS, and innovation studies (Hess and Sovacool 2020). On a generic level, STS as applied to energy studies "[...] *is a research field that provides the capacity to see the interconnections, mutual shaping, co-constitution, or coproduction of the technical, social, and natural*" (Hess and Sovacool 2020, p. 2). The MLP too, focuses on co-construction, but mainly within micro-level niches where "*technology, user preferences, regulation, symbolic meaning, infrastructure, and production systems*" (Geels 2006) are co-constructed. Arguably, however, such co-construction has not been a key focus within the MLP, which has been more concerned with the search for what Geels (2007) calls abstract patterns and explanatory mechanisms. Through this, the MLP has been flagged as having ambitions as "a middle-range theory", situated between grand theory and mundane practice (ibid.).

Hence, while theories are based on similar foundations and have much in common, there is also some tension between the MLP and STS in terms of analytical scope. Where STS-analysts have tended to focus on localized specificities of technology development, the MLP has tended to focus more on generic explanations. We believe that this divide does not only concern a division in how to account for stability and change, but more broadly, there are certain issues that tends to become back-grounded or even black-boxed within the MLP, in part, due to its quest for generic explanations of regime change. Examples are aspects of justice, controversies, practices, politics and power. In other words, while exploring fairly similar phenomena, MLP and STS scholars have tended to ask different questions: while the MLP focus more on the systemic aspects of technological innovation journeys, STS scholars have tended to be more interested in probing broader *consequences* of such journeys.

MLP scholars have engaged with the ontological and epistemological challenges arising from critique that has noted the relative absence of elements like practices and contestation. This has resulted in more refined models and efforts to shift focus and integrate new types of questions in MLP studies (e.g. Geels 2010a, 2011; Vasileiadou and Safarzyńska 2010). Nevertheless, there is still a certain style within MLP scholarship that favours a focus on innovation journeys: stories of how such journeys came about, and more recently, how to accelerate such innovation journeys (e.g. Roberts et al. 2018). We therefore firmly believe that STS have an important role in broadening and deepening the understanding of energy and sustainability transitions, beyond what is currently achieved through MLP and related innovation system focused approaches. To be more specific, while rooted in a socio-technical understanding of reality, studies within the MLP tend to have a techno-centric focus, centred around the introduction of new technologies, or the phasing out of old technologies.

While STS shares many of the same interests in the emergence of new technologies (e.g. Bijker et al. 1987), its focus is often distinctly different from that of MLP scholars. Within STS, there is, for example, a long-standing tradition of studying and normatively promoting processes of public engagement with science and technology (e.g. Wynne 1992; Jasanoff 2012; Chilvers and Kearnes 2015). Aspects such as inclusion, democratization and engagement is rarely made central within studies from the MLP, perhaps because such activities tends to slow down rather than speed up innovation processes (see e.g. Rosa 2013 for a discussion on the relationship between speed of innovation and democratic practice).

Endeavours of studying and advancing public engagement with and democratization of science and technology rest on the assumption that public engagement and participation can be resources for improving the quality of science and technology. Further, it also rests on the assumption that the highly specialized forms of expertise that typically produce cutting edge science and technology, is not always sensitive to the potentially wide societal implications of their own proposals. Finally, such endeavours rest on the assumption that producing and implementing new technologies do not only discretely impact specific sectors, regimes or industries but that they are also essential in shaping our societies more fundamentally.

Thus, while sharing a somewhat common ground when it comes to explaining stability or change as multi actor-processes that are constituted by the alignment of multiple simultaneous processes across society through social and technical means, STS narratives tend to focus attention elsewhere than most MLP-studies. Our ambition here is not to carve out an entire agenda for STS within the study of sustainability transitions, but rather to point briefly to three related trends that inspire us here; and we think STS offers particularly important insights that should be acknowledged and engaged with more broadly by transitions scholars and those interested in energy and sustainability transitions.

First, and most generically, critique from STS and related fields have arguably inspired a turn towards focusing on actors and the relations between actors and technologies in sustainability transition studies. Such critique has illustrated that the production of societal conditions that promote socio-technical novelty is not limited to niche activities. Instead, heterogeneous sets of actors, including incumbent actors, can work to produce social and technological innovation, and also the sites and spaces where technologies are intended to work (Åm 2015; Pallesen and Jenle 2018; Skjølsvold and Ryghaug 2020). Hence, different types of actors can become transition actors (Ryghaug et al. 2018; Sørensen et al. 2018). For us, considerations about the identification of transition actors, but also more abstract patterns of transition activity and agency will be important in discussions about how and why pilot and demonstration projects emerge, as well as in discussions about work to up-scale such pilots.

Second, and keeping with STS' focus on public engagement with science and technology, much work from STS-scholars on energy and sustainability transitions focus on the character of public participation in, engagement with and support for transitions. Technical and economic expertise involved in the production and implementation of new renewable energy technologies, have tended to see the public as a barrier to the successful implementation of their technologies (e.g. Barnett et al. 2012; Skjølsvold 2012; Eaton et al. 2017). STS scholars, on the other hand, have tended to highlight publics as potential resources of innovation (e.g. Lie and Sørensen 1996; Oudshoorn and Pinch 2003), who might contribute not only by accepting technologies a1zssociated with the energy transition, but by shaping the roles of these technologies in society (Ryghaug et al. 2018; Skjølsvold et al. 2018; Throndsen et al. 2017). Hence, participation is a key phenomenon, both as an analytical category, and as a practical resource for realizing energy transition goals.

An important aspect here is that rather than seeing participation as the outcome of individual choice, STS-contributions tends to emphasize the collective production of conditions for participation (Chilvers and Longhurst 2016; Chilvers and Kearnes 2015; Skjølsvold et al. 2018). If innovators see the public as a barrier to implementation, they tend to produce a space where potential modes of participation are to accept and use technologies, or to reject and protest. Through a relational and co-productionist gaze, the responsibility of achieving an inclusive transition becomes distributed to more actors. Technologies can be designed to be inclusive and processes of organization can be conducted in inclusive ways. For us, understanding how such inclusivity can be achieved, and what stands in its way, is a central ambition.

Third and finally, the discussions above suggest that through mobilizing STS in discussions about energy transitions, one can gain a more symmetrical understanding of the social and technical processes of sustainability transitions. This would entail asking questions not only about how to change socio-technical systems, but also to ask more broadly about social aspects of sustainability transitions. On a basic level, social aspects and social categories have also become more prominent in transition studies based on MLP. Here, social categories, however, tends to be analysed directly in relationships to specific technologies. Examples include discussions about the role of technology users in energy transitions (Schot et al. 2016), or on different types of cultural repertoires (e.g. Swedish collectivism, Dutch consensus-based society, and the emphasis on individual freedoms that predominates in the UK) affects the speed of transition (Roberts et al. 2018). While we think such studies greatly enriches transition studies, they implicitly also run the risk of attributing the potential of change to technologies, while casting the social elements of societies as stable. Sustainability transitions, however, will require the transformation of both technology and society, which means that we cannot afford the privilege of stability to either one.

Thus, a focus on particular actors, objects and relations, opens the door to explore generic issues in a different way than through those studies based on typical multi-level understandings. To us, this means a keen interest in understanding changes within politics, lifestyles, cultures and understandings, in and of itself, and not foremost as residual categories that surround technological systems. STS-literatures opens for the analysis of how the social is produced (e.g. Latour 2005), or invented (Marres et al. 2018). Foregrounding the "social" of socio-technical might bring us closer to what Jasanoff (2018) has called "a humble approach to energy futures"—an approach that foregrounds society and issues such as justice, inclusion and equity together with ideas about which sort of societies we want to produce through innovation. We believe that the types of pilot and demonstration projects that we study in this book have significant potential to contribute to such an agenda, but that they currently do so in a relatively limited way. Hence, our contribution here should be read as constructive criticism, which seeks to expand on the logics of contemporary innovation in the name of sustainability.

In what follows, we will shift focus from theory to practice and sketch some key developments within the empirical field we engage with in this book.

The Empirical Field: A Brief Look on Trends and Developments

Considering the climate and sustainability challenges the planet is facing, there is a need to curb emissions from across sectors drastically, and quickly. On the one hand, this entails replacing the carbon-intensive ways that energy is produced for example, from oil, gas and coal to solar-, wind- and bio-energy. On the other hand, it entails changing the way energy is used for example, reducing energy consumption and improving energy efficiency. In other words, there is a need for drastic changes in what has traditionally been described as the production- and the demand-side of energy systems.

Over the last 10 years the developments on the production side of this equation has arguably accelerated drastically. We have seen unprecedented levels of solar and wind power being installed and delivered to the grid; a trend that most actors believe will continue (e.g. IEA 2019). Hence, if we return to the earlier discussed phases of sustainability transitions (Rotmans et al. 2001), many technologies are currently in a phase of acceleration. What does this acceleration entail in practice, and why has piloting and demonstration projects become so central in this period?

A few things are worth noting in relation to these questions. First, renewable energy production sites tend to be smaller than traditional fossil fuel power plants. They are also typically much more distributed throughout the electricity grid—electricity is no longer only generated in a few, large scale facilities, but rather at a range of different sites and scales. A quite common trait, then, is that energy is produced much closer to where energy is consumed. A solar panel can be located on your roof, a wind turbine can be on or close to your own property.

These changes, however, do not only concern size and proximity to production facilities, they also imply changes in the character of energy supply, which becomes more variable with increasing shares of renewable energy technologies. Fossil based power plants can deliver stable electricity loads 24/7, while the production of electricity from sun and wind is volatile and varies with season, time of day and weather. Hence, there is not necessarily a good match between when electricity is produced and when it is in demand. This coincides with another trend in energy systems across Europe, namely that an increasing share of services such as heating, cooling and transportation gradually is becoming electrified. This means that we have a situation where electricity production becomes more variable, while electricity demand is increasing.

These new dynamics clearly illustrates the socio-technical character of the current transition. To succeed in transforming our energy systems in a more sustainable direction, it is not enough to implement new renewable energy technologies, or even to ensure that this happens while coal and other fossil fuels are phased out. The task at hand is a balancing act which involves changing how millions of actors across Europe and beyond, use and relate to electricity on a daily basis. A key term often mobilized to describe this need for a new type of dynamics is *demand side flexibility*. Demand side flexibility entails that actors such as households, small businesses and industry actors should agree to shift some of their electricity demand to periods with less demand to avoid grid congestion problems, to avoid investments towards increasing the capacity of current electricity grids and to facilitate the implementation of variable renewables (e.g. Friis and Christensen 2016).

Currently, many actors hope that smart grids and smart energy technologies will serve as technological enablers of such flexibility (e.g. Torriti 2020), and many of the pilot and demonstration projects we explore in this book come from this domain. Making the grid "smart" entails augmenting energy systems with software, sensors and other ICT technologies, which can enable new forms of communication between actors in the energy system, new pricing schemes and the automation of certain actions (e.g. Silvast et al. 2018; Skjølsvold et al. 2015). Demand response technologies and services are a key element of the smart grid, typically seeking to influence the timing and intensity of energy demand. Common examples include time of use (TOU) tariffs, critical peak pricing, feedback technologies and automated demand controllers (Ingeborgrud et al. 2020; Torriti 2015). Adding complexity to these discussions are new technological developments for instance within battery technology, that suggests batteries from electric vehicles or stand-alone batteries might play an increasingly important role in providing flexibility (see e.g. Noel et al. 2019).

Another characteristic of the developments we have sketched above, is that the technological changes are accompanied by changes in the roles of actors that are involved in the system, or the emergence of entirely new roles. New actors are producing energy: prosumers, energy cooperatives and new types of companies. ICT actors increasingly find opportunities in the smart energy field. New types of business emerge, such as those that specialize in aggregating the flexibility e.g. from fleets of electric vehicles. Many have predicted that the energy industry will see a wave of disruption in which the logic of incumbent infrastructures and industries are replaced by new business models and new forms of organization around the production and distribution of electricity (e.g. Parag and Sovacool 2016). Others point out that the future not only holds new technologies and new roles, but that the character of the very systems might change drastically, for example, becoming microgrids where energy is managed as a common pool resource (Wolsink 2012).

The discussions above point to some developments within energy systems and those should be recognizable throughout Europe and beyond. They also illustrate a generic challenge of energy transitions, which points towards the centrality of the types of projects that we explore in this book.

As energy transitions powered by variable renewable energy production and increasing electrification unfold, analytical and practical complexity increases (Turnheim et al. 2018). Potential roles change, practices change, sectors are coupled in new and sometimes surprising ways, new business models and modes of organization emerge and price structures change. The effects of all these simultaneous changes are difficult to assess even if one only focuses on technologies and looks at the energy system in isolation. Zooming out to contemplate how all of these changes might feed into broader societal processes, it soon becomes clear that we are dealing with a set of open ended and potentially wicked problems (e.g. Buchanan 1992) and as such, many actors see the need to try out and test elements of such ecosystems in localized settings.

These complexities are parts of the backdrop for the current interest in pilot and demonstration projects amongst innovators, policy makers and systems designers. This complexity is also what sparks our interest as researchers. On the one hand, our interest here is fuelled by a curiosity about what it takes to make such socio-technical configurations 'work' in practice. On the other hand, our background in STS and energy social science, allows us to take this interest beyond asking how pilot projects may support and shape technology-oriented innovation and to focus more deeply on how pilot and demonstration projects are made, what their societal and political consequences are, and whether they cater for democratic participation or not. The remainder of this book will be dedicated to these questions.

References

Åm, H. (2015). The sun also rises in Norway: Solar scientists as transition actors. *Environmental Innovation and Societal Transitions, 16*, 142–153.

Barnett, J., Burningham, K., Walker, G., & Cass, N. (2012). Imagined publics and engagement around renewable energy technologies in the UK. *Public Understanding of Science, 21*(1), 36–50.

Bijker, W. E., Hughes, T. P., & Pinch, T. J. (Eds.). (1987). *The social construction of technological systems: New directions in the sociology and history of technology*. MIT Press.

Broto, V. C., & Bulkeley, H. (2013). Maintaining climate change experiments: Urban political ecology and the everyday reconfiguration of urban infrastructure. *International Journal of Urban and Regional Research, 37*(6), 1934–1948.

Buchanan, R. (1992). Wicked problems in design thinking. *Design Issues, 8*(2), 5–21.

Chilvers, J., & Kearnes, M. (Eds.). (2015). *Remaking participation: Science, environment and emergent publics*. Routledge.
Chilvers, J., & Longhurst, N. (2016). Participation in transition (s): Reconceiving public engagements in energy transitions as co-produced, emergent and diverse. *Journal of Environmental Policy & Planning*, *18*(5), 585–607.
Chilvers, J., Pallett, H., & Hargreaves, T. (2018). Ecologies of participation in socio-technical change: The case of energy system transitions. *Energy Research & Social Science*, *42*, 199–210.
Coenen, L., Raven, R., & Verbong, G. (2010). Local niche experimentation in energy transitions: A theoretical and empirical exploration of proximity advantages and disadvantages. *Technology in Society*, *32*(4), 295–302.
Collins, H. M. (1988). Public experiments and displays of virtuosity: The core-set revisited. *Social Studies of Science*, *18*(4), 725–748.
Eaton, W. M., Burnham, M., Hinrichs, C. C., & Selfa, T. (2017). Bioenergy experts and their imagined "obligatory publics" in the United States: Implications for public engagement and participation. *Energy Research & Social Science*, *25*, 65–75.
European Commission. (2020). *Horizon 2020. Work Programme*. Brussels. Accessed from: https://ec.europa.eu/research/participants/data/ref/h2020/wp/2018-2020/main/h2020-wp1820-intro_en.pdf
Frantzeskaki, N., Borgström, S., Gorissen, L., Egermann, M., & Ehnert, F. (2017). Nature-based solutions accelerating urban sustainability transitions in cities: Lessons from Dresden, Genk and Stockholm cities. In *Nature-based solutions to climate change adaptation in urban areas* (pp. 65–88). Cham: Springer.
Friis, F., & Christensen, T. H. (2016). The challenge of time shifting energy demand practices: Insights from Denmark. *Energy Research & Social Science*, *19*, 124–133.
Geels, F. W. (2002). Technological transitions as evolutionary reconfiguration processes: A multi-level perspective and a case-study. *Research Policy*, *31*(8–9), 1257–1274.
Geels, F. W. (2004). From sectoral systems of innovation to socio-technical systems: Insights about dynamics and change from sociology and institutional theory. *Research Policy*, *33*(6–7), 897–920.
Geels, F. W. (2005). The dynamics of transitions in socio-technical systems: A multi-level analysis of the transition pathway from horse-drawn carriages to automobiles (1860–1930). *Technology Analysis & Strategic Management*, *17*(4), 445–476.
Geels, F. W. (2006). Co-evolutionary and multi-level dynamics in transitions: The transformation of aviation systems and the shift from propeller to turbojet (1930–1970). *Technovation*, *26*(9), 999–1016.

Geels, F. W. (2007). Feelings of discontent and the promise of middle range theory for STS: Examples from technology dynamics. *Science, Technology, & Human Values, 32*(6), 627–651.

Geels, F. W. (2010a). The role of cities in technological transitions: Analytical clarifications and historical examples. In H. Bulkeley, V. Castan Broto, M. Hodson, & S. Marvin (Eds.), *Cities and low carbon transitions* (pp. 29–44). Routledge.

Geels, F. W. (2011). The multi-level perspective on sustainability transitions: Responses to seven criticisms. *Environmental Innovation and Societal Transitions, 1*(1), 24–40.

Geels, F. W., & Schot, J. (2007). Typology of sociotechnical transition pathways. *Research Policy, 36*(3), 399–417.

Geels, F. W., Sovacool, B. K., Schwanen, T., & Sorrell, S. (2017). Sociotechnical transitions for deep decarbonization. *Science, 357*(6357), 1242–1244.

Hargadon, A. B., & Douglas, Y. (2001). When innovations meet institutions: Edison and the design of the electric light. *Administrative Science Quarterly, 46*(3), 476–501.

Hargreaves, T., Longhurst, N., & Seyfang, G. (2013). Up, down, round and round: Connecting regimes and practices in innovation for sustainability. *Environment and Planning A, 45*(2), 402–420.

Hess, D. J., & Sovacool, B. K. (2020). Sociotechnical matters: Reviewing and integrating science and technology studies with energy social science. *Energy Research & Social Science, 65*, 101462.

Hughes, T. P. (1993). *Networks of power: Electrification in Western society, 1880–1930*. JHU Press.

Hughes, S., Chu, E. K., & Mason, S. G. (2018). *Climate change in cities*. Cham: Springer.

IEA. (2019). Renewables 2019, IEA, Paris. Retrieved from https://www.iea.org/reports/renewables-2019

Ingeborgrud, L., Heidenreich, S., Ryghaug, M., Skjølsvold, T. M., Foulds, C., Robison, R., ... Mourik, R. (2020). Expanding the scope and implications of energy research: A guide to key themes and concepts from the Social Sciences and Humanities. *Energy Research & Social Science, 63*, 101398.

Jasanoff, S. (2012). *Science and public reason*. Routledge.

Jasanoff, S. (2018). Just transitions: A humble approach to global energy futures. *Energy Research & Social Science, 35*, 11–14.

Latour, B. (1983). Give me a laboratory and I will raise the world. In K. Knorr-Cetina & M. Mulkay (Eds.), *Science observed: Perspectives on the social study of science* (pp. 141–170). London: Sage.

Latour, B. (2005). From realpolitik to dingpolitik. In *Making things public: Atmospheres of democracy* (p. 1444). MIT Press.

Lie, M., & Sørensen, K. H. (Eds.). (1996). *Making technology our own?: Domesticating technology into everyday life*. Scandinavian University Press North America.
Markard, J., & Truffer, B. (2008). Technological innovation systems and the multi-level perspective: Towards an integrated framework. *Research Policy*, *37*(4), 596–615.
Marres, N. (2016). *Material participation: Technology, the environment and everyday publics*. Springer.
Marres, N. (2018). What if nothing happens? Street trials of intelligent cars as experiments in participation. In *TechnoScience in society, sociology of knowledge yearbook* (pp. 1–20). Nijmegen: Springer/Kluwer.
Marres, N. (2020). What if nothing happens? Street trials of intelligent cars as experiments in participation. In *TechnoScienceSociety* (pp. 111–130). Cham: Springer, Chicago.
Marres, N., Guggenheim, M., & Wilkie, A. (2018). *Inventing the social*. Manchester: Mattering Press.
Martiskainen, M., Heiskanen, E., & Speciale, G. (2018). Community energy initiatives to alleviate fuel poverty: The material politics of Energy Cafés. *Local Environment*, *23*(1), 20–35.
Naber, R., Raven, R., Kouw, M., & Dassen, T. (2017). Scaling up sustainable energy innovations. *Energy Policy*, *110*, 342–354.
Noel, L., Rubens, d., G. Z., Kester, J., & Sovacool, B. K. (2019). *Vehicle-to-grid: A Sociotechnical transition beyond electric mobility*. Springer.
Oudshoorn, N. E., & Pinch, T. (2003). *How users matter: The co-construction of users and technologies*. MIT Press.
Pallesen, T., & Jenle, R. P. (2018). Organizing consumers for a decarbonized electricity system: calculative agencies and user scripts in a Danish demonstration project. *Energy Research & Social Science*, 38, 102–109.
Parag, Y., & Sovacool, B. K. (2016). Electricity market design for the prosumer era. *Nature Energy*, *1*(4), 1–6.
Raven, R., Kern, F., Verhees, B., & Smith, A. (2016). Niche construction and empowerment through socio-political work. A meta-analysis of six low-carbon technology cases. *Environmental Innovation and Societal Transitions*, *18*, 164–180.
Roberts, C., Geels, F. W., Lockwood, M., Newell, P., Schmitz, H., Turnheim, B., & Jordan, A. (2018). The politics of accelerating low-carbon transitions: Towards a new research agenda. *Energy Research & Social Science*, *44*, 304–311.
Rogelj, J., Shindell, D., Jiang, K., Fifita, S., Forster, P., Ginzburg, V. … Vilariño, M. V. (2018). Mitigation pathways compatible with 1.5°C in the context of sustainable development. In: *Global Warming of 1.5°C. An IPCC Special Report on the impacts of global warming of 1.5°C above pre-industrial levels and related global greenhouse gas emission pathways, in the context of strengthening the*

global response to the threat of climate change, sustainable development, and efforts to eradicate poverty. Retrieved May 26, 2020, from https://www.ipcc.ch/site/assets/uploads/sites/2/2019/02/SR15_Chapter2_Low_Res.pdf

Rosa, H. (2013). *Social acceleration: A new theory of modernity*. Columbia University Press.

Rotmans, J., Kemp, R., & Van Asselt, M. (2001). More evolution than revolution: Transition management in public policy. *Foresight-The Journal of Future Studies, Strategic Thinking and Policy, 3*(1), 15–31.

Ryghaug, M., Ornetzeder, M., Skjølsvold, T. M., & Throndsen, W. (2019). The role of experiments and demonstration projects in efforts of upscaling: an analysis of two projects attempting to reconfigure production and consumption in energy and mobility. *Sustainability, 11*(20), 5771.

Ryghaug, M., Skjølsvold, T. M., & Heidenreich, S. (2018). Creating energy citizenship through material participation. *Social Studies of Science, 48*(2), 283–303.

Sadowski, J., & Levenda, A. M. (2020). The anti-politics of smart energy regimes. *Political Geography, 81*, 102202.

Schot, J., & Kanger, L. (2018). Deep transitions: Emergence, acceleration, stabilization and directionality. *Research Policy, 47*(6), 1045–1059.

Schot, J., Kanger, L., & Verbong, G. (2016). The roles of users in shaping transitions to new energy systems. *Nature Energy, 1*(5), 1–7.

Silvast, A., Williams, R., Hyysalo, S., Rommetveit, K., & Raab, C. (2018). Who 'uses' smart grids? The evolving nature of user representations in layered infrastructures. *Sustainability, 10*(10), 3738.

Skjølsvold, T. M. (2012). Publics in the pipeline. In N. Möllers & K. Zachmann (Eds.), *Past and present energy societies*. Bielefeld: Transcript Verlag.

Skjølsvold, T. M., & Ryghaug, M. (2020). Temporal echoes and cross-geography policy effects: Multiple levels of transition governance and the electric vehicle breakthrough. *Environmental Innovation and Societal Transitions, 35*, 232–240.

Skjølsvold, T. M., Ryghaug, M., & Berker, T. (2015). A traveler's guide to smart grids and the social sciences. *Energy Research & Social Science, 9*, 1–8.

Skjølsvold, T. M., Throndsen, W., Ryghaug, M., Fjellså, I. F., & Koksvik, G. H. (2018). Orchestrating households as collectives of participation in the distributed energy transition: New empirical and conceptual insights. *Energy Research & Social Science, 46*, 252–261.

Smith, A., & Raven, R. (2012). What is protective space? Reconsidering niches in transitions to sustainability. *Research Policy, 41*(6), 1025–1036.

Sørensen, K. H., Lagesen, V. A., & Hojem, T. S. M. (2018). Articulations of mundane transition work among consulting engineers. *Environmental Innovation and Societal Transitions, 28*, 70–78.

Szulecki, K. (2018). Conceptualizing energy democracy. *Environmental Politics, 27*(1), 21–41.

Throndsen, W., & Ryghaug, M. (2015). Material participation and the smart grid: Exploring different modes of articulation. *Energy Research & Social Science, 9*, 157–165.

Throndsen, W., Skjølsvold, T. M., Ryghaug, M., & Christensen, T. H. (2017). From consumer to prosumer. Enrolling users into a Norwegian PV pilot. ECEEE Summer Study Proceedings, 2017.

Torriti, J. (2015). *Peak energy demand and demand side response*. Chicago: Routledge.

Torriti, J. (2020). *Appraising the economics of smart meters: Costs and benefits*. Routledge.

Turnheim, B., Wesseling, J., Truffer, B., Rohracher, H., Carvalho, L., & Binder, C. (2018). Challenges ahead: Understanding, assessing, anticipating and governing foreseeable societal tensions to support accelerated low-carbon transitions in Europe. In *Advancing energy policy* (pp. 145–161). Cham: Palgrave Pivot.

Vasileiadou, E., & Safarzyńska, K. (2010). Transitions: Taking complexity seriously. *Futures, 42*(10), 1176–1186.

Wilkie, A., & Michael, M. (2018). Designing and doing: Enacting energy-and-community. In N. Marres, M. Guggenheim, & A. Wilkie (Eds.), *Inventing the social* (pp. 125–148). Manchester: Mattering Press.

Wolsink, M. (2012). The research agenda on social acceptance of distributed generation in smart grids: Renewable as common pool resources. *Renewable and Sustainable Energy Reviews, 16*(1), 822–835.

Wynne, B. (1992). Misunderstood misunderstanding: Social identities and public uptake of science. *Public Understanding of Science, 1*(3), 281–304.

Open Access This chapter is licensed under the terms of the Creative Commons Attribution 4.0 International License (http://creativecommons.org/licenses/by/4.0/), which permits use, sharing, adaptation, distribution and reproduction in any medium or format, as long as you give appropriate credit to the original author(s) and the source, provide a link to the Creative Commons licence and indicate if changes were made.

The images or other third party material in this chapter are included in the chapter's Creative Commons licence, unless indicated otherwise in a credit line to the material. If material is not included in the chapter's Creative Commons licence and your intended use is not permitted by statutory regulation or exceeds the permitted use, you will need to obtain permission directly from the copyright holder.

CHAPTER 2

The Co-production of Pilot Projects and Society

Abstract This chapter discusses the shaping of pilot projects. Against a critique that such projects tend to be shaped top-down by powerful actors, our discussion notes how such projects are also shaped locally by materiality, culture, actors, interests and issues. Through this we show how projects end up looking very different from each other while enacting diverse socio-technical futures. We discuss three types of pilot projects: technology-oriented projects, geographically bound projects and national laboratories. We argue that pilot projects, in either form tend to mirror and amplify the interests of involved actors, and we proceed to discuss the potential politics of such projects. We do this by discussing processes of scaling up pilot projects, and through upscaling, shaping broader aspects of society. As these projects often have wide transformational ambitions, we conclude that a focus on who participates and who does not is central for future research.

Keywords Co-production • Interests • Pilot projects • Upscaling

Introduction: Why Study Pilot Projects?

One of our goals in this book is to develop a broad understanding of the roles that pilot and demonstration projects play in unfolding sustainability transitions in contemporary societies. Building on insights from science and technology studies (STS) we have an interest in questions about how specific historical and social dynamics shape technologies (e.g. Williams and Edge 1996), and how technologies are constructed through the interpretations and work of different social groups (Pinch and Bijker 1984). Within such a perspective, pilot and demonstration projects that seek to advance smart grids or electro mobility, could be interpreted as distinct processes and products shaped by and embedded within the societies where they emerge. The situation is more complex, however, because neither society, nor technologies are static but change over time. One explanation for this is that technologies also contribute to shaping the very societies they become part of (e.g. Latour 1987). Hence, the pilot and demonstration projects we study are both products of and produces society—they are in other words co-produced (Jasanoff 2004).

Such an entry point to the study of pilot and demonstration projects in not only semantic word play. The perspective also underscores the importance of scrutinizing the activities within, as well as the outcomes, of such projects, beyond the notion of innovation journeys. Working from the assumption that new technologies are essential to future social orders, it becomes important to probe both how pilot projects emerge, and also which social orders they explicitly or implicitly promote. In asking such questions, we follow a long tradition of STS scholars working to understand the politics of technologies (Winner 1980; Sørensen 2004; Nahuis and Van Lente 2008).

Studying the shaping or construction of pilot projects is important, in part because it might shed light on whose politics such pilots enact. Critical readings of pilot projects within new and smart energy technologies tend to highlight their emergence as a top-down phenomenon (Throndsen 2017), shaped primarily by technology policies that stimulate experimentation in the interests of powerful actors. As an example, much experimentation within smart energy over the last 10–15 years can be described as emerging as a form of response to national or European policy agendas pushing to implement smart grids (e.g. Ballo 2015; Vesnic-Alujevic et al. 2016). European policy goals are also operationalized through priorities in funding bodies such as the Horizon 2020 (see Chap. 4). Such funding steers and shape innovation activities in distinct directions by circulating capital, ideas and technologies (see e.g. Skjølsvold et al. 2020; Rosenow and Kern 2017).

However, the image of the smart grid as emerging primarily top-down is not entirely clear-cut. As one zooms in on specific demonstration and pilot projects, these tend to be made up also by highly localized networks of actors who do not necessarily see themselves as responding to top-down technology push. Rather, they engage in issues that are constituted locally, and often rooted in quite specific problems, shaped by material conditions such as the local configuration of the electricity grid, distinct social issues or economic conditions (Skjølsvold and Ryghaug 2015). In this chapter we are interested in exploring how pilot and demonstration projects are constituted through diverse relations that do not necessarily sit well within the categories of bottom-up or top-down.

When zooming in on the techno-politics of pilot and demonstration projects related to renewable energy, smart grids and electromobility, we are confronted with an interesting duality: On the one hand, their politics are often quite explicit. If we mobilize the language from the multi-level perspective on socio-technical transitions as discussed in Chap. 1 (Geels 2002); they often seek to challenge, change or replace existing socio-technical regimes. Following these explicit political agendas of many pilot and demonstration projects leads us towards an interest in how such projects, and the actors that take part, work to expand on, or to scale-up their proposed solutions (e.g. Naber et al. 2017; Ryghaug et al. 2019).

On the other hand, the delegation of the production of future social order to pilots, demonstrations and experiments also leads to discussions about their more implicit politics. Many scholars have argued that innovation activities in the name of sustainability, such as those discussed here, signals the emergence of a form of post-politics (Rosol et al. 2017) or anti-politics (Sadowski and Levenda 2020), where democratic processes are suspended at the expense of expert assessment and corporate interests. Others again, are more hopeful and highlight the political potential of such projects as sites that might produce new forms of democratic participation (Marres 2013, 2016). To us, there is no determinism attached to pilot and demonstration projects that leads to one or the other of these two positions. Instead, this tension signals the importance of studying the inner workings of such innovation activities, because so much is at stake within and around them: they are places where politics, participation and society might be steered in different directions.

The remainder of this chapter will be structured along the lines of the discussion sketched out above. First, we will discuss the societal shaping and emergence of pilot projects. While pilots might often appear small, they are characterized by the mobilization of a wide repertoire of resources and rationales that shape their content and activity. Building from this we move on to discuss these projects as political endeavours and how they in turn might shape societies; how they might scale up, what scaling up might entail, and the politics of pilot and demonstration projects more broadly.

Exploring the Shaping of Pilot Projects

Pilot and demonstration projects within energy and mobility are diverse and are developed by a highly diverging set of actor constellations with different rationales. In what follows, we will mobilize a sociologically oriented sensitivity from STS (e.g. Williams and Edge 1996; Pinch and Bijker 1984), as we discuss the characteristics of some typical forms of pilot projects. In our discussion, we move from strongly focused technology-oriented projects that are confined to relatively small spaces to pilots that are geographically much broader. We cover the following types of projects:

- The technology-oriented trial
- The geographically bound pilot
- The national laboratory

Through this exercise, we can explore pilots and demonstrations at different scales, but just as importantly, we probe their differences in terms of focus. While all pilot projects are socio-technical in nature, there are important differences in the ways that projects are set up: Some are primarily interested in harvesting technological or theoretical insights, while others more actively seek to explore relationships between technologies and social change. Between different types of projects, also the types of resources involved differ greatly, as do the types of interests and actor constellations involved in them. To us, this illustrates that it matters where and by whom pilot and demonstration projects are set up, because this is a key aspect of the politics that, in turn, become important in articulating future social order (Nahuis and Van Lente 2008). As we will discuss later in this book, this observation also places great responsibilities on the shoulders of all actors who fund, promote and enact such projects: the ways this type of innovation and research activities are configured might shape our collective futures in important ways.

Technology-Oriented Trials: From Laboratories to Regional Specificities

Pilot and demonstration projects have become a distinct way of testing new solutions and instigating socio-technical change in areas in need of transformation or in areas where actors have ambitions for technological advancement and innovation. Smart grids and smart energy technologies are important examples, as areas where countless pilot and demonstration activities have been set up all over Europe and beyond during the last years. As an example, the database of smart grid projects across the European Union (EU) Member States names over 950 smart grid pilot studies conducted between 2010 and 2017 totalling around 5 billion Euros in investments (Gangale et al. 2017).

In short, smart grids entail augmenting electricity grids with software, sensors and new forms of controls, which are expected to result in much more active management of the resources flowing through the grid, on behalf of traditional energy producers, electricity grid operators and electricity consumers (e.g. Silvast et al. 2018; Goulden et al. 2014; Skjølsvold et al. 2015). These technologies are, on the one hand, expected to play a prominent role in enabling wide-scale integration of variable renewable energy production and to handle peak electricity demand in strained electricity grids. On the other hand, such technologies are also expected to enable the emergence of new types of market actors and market structures, who for example, capitalize on aggregating and commoditizing the choices of countless consumers and prosumers (e.g. Curtis et al. 2018).

These technologies have comprised a central element in energy transition roadmaps in places as diverse as the US, the UK and China over the last decade (Berker and Throndsen 2017), emphasizing our statement about smart grids being strongly promoted through top-down policies. Currently, the European Technology and Innovation Platform "Smart networks for energy transition" (SNET) argues in its "Vision 2050", that smart and digital technologies are one of the key building blocks of the energy transition. This platform is set up as an instrument to guide the research and innovation priorities of the European Commission, and as such, it represents an important voice both in terms of defining key policy priorities and in terms of funnelling financial resources into those priorities. The report highlights:

In 2050, digitalisation facilitates services and the full integration of all kinds of energy systems: Several million households actively participate in real-time, automated demand response (electricity, heating and cooling) with connected appliances and equipment, in addition to the existing and emerging solutions for industry and commerce. (ETIP SNET 2018)

Further, the report goes on to highlight that: "Aggregation of smart charging technologies for electric vehicles, stationary batteries, heat pumps and power-to-gas provides controllable electricity loads". This is only one, of countless future smart energy visions, which provide direction for innovation, which in this case is also strongly linked to the provision of potential monetary and intellectual resources.

In a Norwegian setting, where we have conducted the main part of our empirical work, much inspiration for smart grid pilot and demonstration activities were triggered by a regulation passed by the Norwegian directorate of water resources and energy (NVE) in 2011. This regulation demanded that all households should have smart meters installed by 2019. In Norwegian policy debates, this implementation has been considered a key stepping stone towards a blanket smart grid implementation, where the underlying rationale was based on socio-economic assumptions about smart meters enabling increased efficiency in electricity markets, rationality in grid operation, grid optimization and active consumption (Ballo 2015; Frøysnes 2014; Skjølsvold 2014). From the point of the authorities, the regulation was also framed as a potential trigger of innovation and industry development at the intersection of industries such as information and communication technology, electricity and construction.

A Smart Grid Laboratory for the Purely Technical?

The above paragraphs indicate that pilot and demonstration projects emerge in a broader context of national and international vision production, as well as regulations, policies and instruments that make resources available for trials with technologies that are thought to enable the materialization of such visions. In Norway, one could observe a surge of smart grid demonstration projects following the formulation of the regulation, where industry actors experimented with new arrangements based on smart electricity meters combined with other technologies under what they perceived to be relatively realistic conditions (Christensen et al. 2013; Fosso et al. 2014; Skjølsvold and Ryghaug 2015).

Most of these trials and pilots have been strongly geared towards advancing technological innovation. In some instances, the activities have been framed by the involved actors as "purely technical". An example of this can be found in the Norwegian Smart Grid Laboratory, which is a physical set-up located on the campus of the Norwegian University of Science and Technology. The laboratory consists of a miniature physical power system. Much of the work done in the laboratory entails researchers and industry actors testing how new technologies respond under technically realistic, but highly controlled circumstances. The key goal, according to the management of the laboratory, is to "research and verify technology and functionality". In other instances, the smart grid laboratory has been framed as a system-oriented infrastructure for pilots and demo projects.

The production and shaping of this laboratory reflect Norwegian research priorities with respect to smart grids in the early 2010s. The laboratory was established with support of close to 30 million NOK (approximately 3M Euros) from the Norwegian Research Council and is institutionally anchored at the intersection of the largest technical university in Norway, and SINTEF, a large, industry-oriented research institution. Further, the establishment of the laboratory was strongly supported by dozens of actors in the electricity grid and production sector, who at the time were facing great uncertainty with respect to how a smart energy future would look. The laboratory enrolled actors from these sectors, in part by appealing to the costs of innovation and of rolling out technological solutions that subsequently fail during operation. Part of the rationale for building the laboratory was also its potential ability to attract further funding, for example, from bodies such as the Horizon 2020 and the Norwegian research council.

While involved actors tends to frame the goals of the laboratory narrowly in terms of developing and testing the functionality of new technologies, an examination of the laboratory with a socio-technical gaze illustrates that it in-fact has a much broader scope. It is a laboratory that mediates between industry needs and research interests, and its activities are shaped by a combination of political priorities both nationally and in the EU, future visions, academic interests, funding mechanisms and industry interests. During its operation, the laboratory has also been shaped further by the needs of the university, who have used the laboratory as a site of public and political engagement, welcoming a steady stream of delegations of energy policy makers and industry leaders that are guided through the facilities to illustrate the work done to materialize smart energy futures. Thus, we clearly see the way that the laboratory, which is supposedly a site of "the purely technical", has taken on the role as the place where the abstract idea of an energy transition can be put on material display to facilitate encounters between high level management of the university, policy makers and industry leaders, as well as between researchers and prospective societal partners. From an STS perspective, this is hardly surprising. While laboratories discursively are often praised as completely de-contextualized or "placeless" places (Kohler 2002), empirical studies of what goes on inside and around laboratories have illustrated the importance of their social, cultural and economic embedding (Latour and Woolgar 1979; Knorr-Cetina 1995). To us, the shaping of this laboratory as a key site of the energy transition in Norway is symptomatic of how a technical university and its epistemic culture, European funding mechanisms and national innovation policies have resulted in a narrow focus on technology development, which is often largely disinterested in how technologies become embedded in societies (Sørensen 2013).

The actors involved in this laboratory and the research, trials and developments within it, publicly reproduces the notion that technology development and testing are distinctly separate activities from making technologies work in society. As an information website for the laboratory highlights:

> Topics such as market solutions, customer behavior and business plans are not covered in the lab but must be done in projects with customers involved that are managed by the energy and grid companies.

To us, this illustrates that the way pilot and demonstration projects are shaped, constructed and framed, are key to also shaping their politics. In this case, an overtly focus on technology has constituted several research and development communities, which if we mobilize the language of the multi-level perspective (Geels 2002) might represent a research regime and a selection environment that explicitly do not engage in social issues. Given the socio-technical complexities of energy transitions, this is problematic, and the consequences of such moves will be discussed later in this book. In what follows, we will concentrate on a set of technology-oriented pilot projects that precisely seeks to explore this relationship between new technologies and their use.

Technology Trials Outside the Laboratory

Beyond the laboratory discussed above, several pilot projects emerged throughout Norway in the same period, all in the wake of the new national regulation that demanded smart meters to be rolled out across the country. Actors within the energy, electricity and building sector have traditionally been conservative, in the sense that they have favoured well proven concepts and materials at low costs, often developing new solutions and business models only when pushed in this direction through regulation (Ryghaug and Sørensen 2009). Hence, a policy-push explanation for the emergence of the pilots is tempting.

However, these first pilot and demonstration projects illustrate that a narrative of policy-push only reveals half the story. The regulation itself was relatively open-ended which meant that it left, in the hands of the grid companies, to make the most of the situation in terms of finding innovative and lucrative solutions that could benefit companies, users and society. In practice, these pilots and demonstration projects were shaped by a set of very diverse and local actor-constellations, which formulated a set of highly different issues and problems to address through smart energy piloting. Thus, smart grid pilots typically have been constructed by local actors across different sectors, who make the pilots parts of pre-existing local realities in quite different ways, while also mobilizing resources and interests from national, and international domains (Skjølsvold and Ryghaug 2015). We will give some examples below.

The High-tech Neighbourhoods of the Smart Grid

Smart energy pilot projects are constituted differently, and consequently, take on very different shapes. Consequently, pilot projects enable the articulation of quite different socio-technical realities. Two illustrative examples will help us see how this might unfold in practice. Our first example is in the south of Norway, a few kilometres outside of the small town, Arendal. Here, the interest in experimenting with smart energy technologies emerged from an alliance between a local electricity production company and actors in the building and property development industry. The initial background was a plan by a construction company to build a new neighbourhood of so-called "plus-houses". These houses should qualify as passive houses, but also be fully equipped with solar panels for prosumption. Geothermal boreholes and solar collectors produced hot water for space heating and showering, and for hot-fill washing machines and dishwashers. The houses were equipped with smart meters and smart home technology, and state of the art ventilation systems. Hence, almost every imaginable technology intended for a smart energy future was tried out. Direction to this work was provided by national and international research and innovation funding.

The projects outspoken aim was to produce technological and practical insights on how such a combination of technologies could work in practice. As such, their approach to innovation differed from that of the laboratory discussed above. Their innovation approach was described as a "living lab", emphasizing the importance of developing technologies that will function when they are to be implemented in society (Haugland forthcoming). When finished, the houses were to be sold on the ordinary retail market for a considerable premium price, targeting early adopters of new "green" technology. The plan for this neighbourhood was eventually becoming a real-life laboratory or a sort of natural experiment for the continued exploration of the relationship between new technologies and energy related practices.

The above story, in a rather straightforward way, illustrates how the interests of different actors' feed into project goals, thereby re-iterating the well-established point of STS that the involvement of different social groups influences technology design (Pinch and Bijker 1984). The involvement of a construction company that also develops and sells properties, resulted in an interest in exploring life in future homes, willingness to pay and new construction techniques. The involved energy companies were interested in how such high-tech living would influence the grid. Primarily, they were concerned with the combination of power intensive technologies and automation, and whether implementing many of these power consuming technologies at the same time and in the same place might lead to unintended consequences such as increased or new kinds of peak load problems.

While the primary interest of the trial was technological, it enabled the mobilization of a distinct focus on social aspects of the technologies tested such as technology—user experiences, thinking about how the new configurations of technologies would affect the identities of the neighbourhoods, and imagining who would want to live in these houses. In doing so, the actors who developed the neighbourhood, the engineers and housing developers, imagined what Strengers (2013, 2014) has named a "resource man"; an energy-interested optimizer who loves new gadgets.

Further, the involved actors cultivated a certain interest in the day-to-day living with all these new technologies beyond how they would impact the grid.

This set the project apart from many smart grid trials, as much of the technical discourse surrounding smart grids often make jumps from abstract concepts like end-user flexibility to the idea of load-shifting or shaving peaks, without discussing the practices that make up electricity consumption (Katzeff and Wangel 2015). This demonstration project, on the contrary, attempted to address the production of flexibility, including the uncertainties that emerged in encounters between new technologies and their users. In sum, the socio-technical repertoire of such a pilot may be describes as being much broader than that of the smart grid laboratory. This outcome can in turn be attributed to a broader set of interests being involved in the formulation of what the smart grid might be, as demonstrated in this example. Our next example invokes an even broader anchoring of different interests and consequently, what may be the focus of a smart grids pilot.

The Smart Grid Shaped by Healthcare Actors

Our second pilot example is located in Stavanger, a city on the west coast of Norway. The city has been expanding rapidly for many years, mainly because it hosts many offshore industries involved in Norwegian oil and gas activities. This has generated a lot of wealth and many well-paid employment opportunities locally, resulting in a substantial population growth in the region and an increased pressure on the electricity grid of the area. Local grid operators and electricity producers thus faced significant challenges in terms of energy provision and security. Expanding the grid capacity, however, was not considered a viable option and local actors considered smart energy innovations to be a more viable path forward.

In doing so, the actors in this region approached the relationship between social and technological development in a more open way than the two cases we have discussed this far. The energy and grid company of the area first hosted a series of workshops, in which the goal was to identify social issues that could be addressed through the mobilization of 'smart'. As such, the involved actors quite explicitly formulated ideas about potential material political strategies. Several actors from local health care institutions participated in the workshops, which served not only as discussions about energy, but about regional development more broadly. A recurring theme which interested the healthcare organizations was what was perceived as a forthcoming demographic transition, in which the share of elderly people in the region would increase drastically. The initial workshops resulted in an alliance between traditional energy sector actors, ICT actors and actors within healthcare.

For the healthcare sector, smart energy technologies represented not only an opportunity to reduce energy consumption and shift loads, but a technological way to tackle their own peak load problems, in the form of crowded healthcare institutions. Together with the other actors, the idea of producing simple home control and automation technologies for the elderly, or for disabled persons as an integral element of the smart grid trial was forged. Such technologies, was the ambition, would make it easier for senior citizens and disabled persons to live in their own private homes, rather than in a healthcare facility. The prospects of transforming energy consumption patterns through using the same home control and automation technologies were seen as a possible added value.

This case illustrates once more that technological outcomes differ between sites, especially when the involved actor groups are divergent. In retrospect, one can be tempted to explain such outcomes by pointing to social and cultural factors such as local demographics and wealth. However, the example also points to the importance of the way such innovation processes are governed. Innovation outcomes do not only depend on where, by whom and when they are conducted. The practices and the work of the involved actors are essential (Callon 1984). In this last example we see how an active and open approach to innovating allowed for sensitizing of the involved actors to a potential link between so-called welfare technologies and smart grid technologies. This also resulted in an explicit criticism of many of the user representations mobilized in many other smart grid projects as this particular project were cast as a necessary move away from typical design processes where one was designing for resource man (Strengers 2013), and a quest to design technologies that everyone, even elderly, could and would use.

Another consequence of this move was that the sphere of influence of smart energy technology pilots expanded, and that their implicit politics would also potentially affect the world of healthcare. On the one hand, we can read this as a democratic opening for the participation of healthcare workers in energy transition activities: this actor group was now engaged in formulating new issues and new politics of energy transition. On the other hand, scholars have noted how welfare technologies often re-distribute health responsibilities from welfare states to other industries and from healthcare workers to the elderly themselves (e.g. Tøndel and Seibt 2019). With such and interpretation, this smart grid pilot project became enrolled in a form of politics which are also central for shaping the future of the healthcare sector and the logics of the welfare state more broadly. Either way, the example illustrates that as increasingly diverse sets of actors with various interests get involved in such innovation endeavours, the number and complexity of potential social questions increase.

In what follows, we will discuss pilot and demonstration projects of a different character and where the local or geographical characteristics have become even more prominent, namely pilots that embrace experimentation within a confined geographical area.

Geographically Bound Pilots

While the projects discussed above test technologies within a laboratory or within a neighbourhood consisting of a few houses, other pilot and demonstration projects focus on a particular geographical area, with the ambition of transforming the whole area over time. The most emblematic version of this is perhaps the city, which has for long been recognized as a site of experimental governance where civil society, commercial actors, municipalities and researchers work together through projects, often under the banner of smart cities. Castan Broto and Bulkeley (2013) identified no less than 627 urban climate change experiments across 100 cities globally, seeking to transform elements such as energy infrastructures, transportation and the built environment in cities. The politics of this development has been subject to criticism, which notes how smart cities have tended to promote a neoliberal agenda, celebrating entrepreneurship and privatizing ethics, while rejecting social justice as a legitimate goal of public policy (Morozov and Bria 2018). In such accounts, 'smart' is most often assumed to be a top-down transformative process.

Other scholars have noted that this is not the whole story. They have highlighted that smart cities can be both situated and purposive, mobilizing local resources and resources that circulate through global networks to shape individual projects and the city in diverse ways (Bulkeley et al. 2014). On the one hand, such differences in analysis might signal the different perspectives of analysts. On the other hand, however, it also highlights the importance of moving beyond emblematic labels such as "smart" to study how distinct innovation processes are shaped and enacted. To us, it highlights that *who* are involved in producing pilot and demonstration projects matters, and the fact that the outcomes and politics of such projects are not pre-determined. In the next section, we will look at another popular geographically bound site that is often preferred for experimenting with smart grids.

The Island as an Example

Cities are not the only geographical units that have become emblematic of pilot and demonstration projects rooted in smart energy technologies. Over the last years, many islands have taken on roles as eco-islands, smart islands or renewable islands. In such instances, the very islands involved in piloting serve as a key resource for the innovation activities. Islands tend to be relatively small and physically separate from mainland geographies, and therefore offer traits that resemble the laboratory. Grydehøj and Kelman (2017) have noted that such traits might allow innovators working on islands to gain quick, but credible results. Transforming the way energy is produced and consumed on an entire island serves as a material and practical illustration of how such a transformation might look in society, more broadly. Such islands, then, are often thought of, not only as technology test sites, but also as models or blueprints for how to achieve society-wide transition at a later stage (Skjølsvold et al. 2020).

Considering the above observations, pilot islands are not merely curious examples of fringe innovation activities, but potential spearheads in the promotion of new social orders rooted in smart energy technologies. As the examples discussed earlier in this chapter, demonstration activities on islands have been heavily promoted by the European Union and its innovation and research program Horizon 2020, which explicitly identifies the 2200 islands of Europe as laboratories and pipelines that are vital for realizing the continents energy transition. Islands, the research and innovation program suggests, are small, vulnerable and dependant on global energy commodity chains, but can be transformed and empowered though innovation, technology and competence provided through Horizon 2020 projects (European Commission 2020).

The narrative of this innovation policy undeniably reads as an agenda of powerful actors pushing technology and innovation from the top down. Island innovators tend to mobilize resources such as those made available through international, national and regional funding bodies in their endeavours, and through this, the activities on islands are shaped by national and international innovation policy. However, this narrative should also be nuanced, as many island pilot and demonstration projects are also articulated around local issues, and enacted by local actors.

The Danish island of Samsø can serve as an example. For more than two decades, this island has been heavily engaged in a series of energy transition and innovation projects, where renewable energy, smart energy and new transportation technologies have been central. Today, many of the activities are operationalized though a large research innovation project funded through the Horizon 2020. The actors involved in the project consist of energy producers and grid companies, ICT companies, as well as various researchers, primarily rooted in the technical sciences. The project's narrative of Samsø reflects the earlier discussed European innovation policy, highlighting that Samsø is small, has renewable resources and needs more efficient energy system management.

However, contemporary innovators on Samsø, also mobilize another element when highlighting why Samsø is an ideal site for smart grid innovation activities. This element is what some scholars call 'an imagined public' (Barnett et al. 2012; Ryghaug and Toftaker 2016; Solbu 2018), an ideal typical representation of a public produced by technology implementers and innovators. Actors who work to implement new technologies often conflate publics either to mean "consumers" of services (Cotton and Devine-Wright 2012) or "barriers" to implementation (Heidenreich 2015; Throndsen 2017), but on Samsø, the public is imagined differently. Here, innovators portray an enthusiastic and communally oriented public, which is understood to be an important enabler of innovation (Skjølsvold et al. 2020).

Compared with the three earlier cases discussed in this chapter, contemporary innovators on Samsø embrace what they perceive as a set of social characteristics of the place and cite these characteristics as essential to succeeding with their pilot. Papazu (2016, 2018) notes how this imagination of Samsø's public is a historical construct, rooted in events that unfolded from 1997 to 2007. During this decade, Samsø became largely self-sufficient with renewable energy through a much-discussed model of community building, public participation and shared ownership. In subsequent years, this narrative about Samsø's communal energy transition has become fixed to the point where it has taken on close to mythical proportions.

The transition story of Samsø has become exported and globalized. Papazu (2018) notes how this is problematic, because it obscures the many socio-material challenges that characterized the transition that unfolded on the Island. To us, the globalization of this narrative also serves to shed light on the shaping of contemporary pilot activities on Samsø. While European innovation policies shape activities on Samsø, part of Samsø's attraction for European funders is likely to also be shaped by its reputation as a communally oriented transition-island. In this way, the narrative becomes re-articulated and re-produced by new policy and funding body actors on a regular basis, which in turn re-establishes Samsø as a lucrative place for future pilot and demonstration activities.

Documents from contemporary pilot activities on Samsø, point towards a complex relationship between the way local interests and the interests of European policy, industry and research shapes these activities. At times, innovation activities on the island are highlighted to be concerned with the creation of local jobs, increasing the population and building new forms of community. At other times, the innovation activities are framed primarily as geared towards testing the future solutions of the European energy transition. Rather than seeing Samsø as the victim of top-down policies to materialize European visions of a smart, renewable and distributed energy future, we interpret this situation as a signal that the politics of pilot and demonstration projects are multiple.

While Papazu (2018) could be right in problematizing the way that publics on Samsø are imagined, we believe there is significant potential in mobilizing ideals about community and kinship as basis for smart energy innovations. Today, however, it is difficult to see how or if these ideals are reflected in the technologies currently tested on Samsø. Grydehøj and Kelman (2017) notes that islands who engage heavily in such innovation activities tends to mobilize terms such as "community" in order to masquerade that they engage in a form of conspicuous and highly competitive pursuit of technology oriented and dominated (sustainability) projects. Paradoxically, this would hardly be sustainable, in the sense that all islands cannot plausibly compete for and be high-tech frontrunner green islands. Further, it stands to reason that exporting and upscaling such a model off-islands would be difficult.

In the above discussion we have moved from pilot projects undertaken in single laboratories, via trials in society to relatively large but geographically bound pilots, illustrated by the island as an example. In what follows, we will make a small conceptual leap, as we will discuss trials at a national scale as pilot projects.

The National Pilot-Project

Trials at this size are quite rare, but not unheard of, and they offer a range of opportunities. Entrenching a pilot project deeply within society in this way means that they encounter all the various actors, organizations and technologies of that society. This means that such pilots will be less "controllable" than for example, earlier discussed examples of pilots in laboratories or pilots limited technology trials. Nevertheless, a country can represent a smaller, less-expensive and faster-responding trial than a broad effort to transform for example, global transport systems. In the past, Iceland has for instance been highlighted as an example of a country-size laboratory for the hydrogen economy (Park 2011). An important point is that such pilots might offer valuable opportunities not only for large-scale technology trials, but also for explorations of the impact of new forms of regulations and policies. In what follows, we will look more closely at one example, namely how one can understand the Norwegian electric vehicle (EV) transition as a national pilot project to develop, promote and implement electromobility, and what such an interpretation might entail.

The Norwegian Case of Electromobility

At the time of writing this book, auto manufacturers around the world are quickly embracing electro mobility and especially electric person cars, and several national and local markets for example, in California and Germany are growing rapidly. Nevertheless, Norway stands out in discussions around electromobility, because the country has promoted the emergence of EV markets in a systematic way for at least two decades, to the extent that that the country indeed, can be described as a national laboratory for electromobility.

The shaping of this laboratory can be described by distinguishing between two phases of development. If we mobilize the language from the multi-level perspective (see Chap. 1), the first period (1990–2009) can be described as a technology niche creation phase (Ryghaug and Skjølsvold 2019). Here, the political goal was primarily to nurture a Norwegian EV industry, and to create a domestic market for this industry. In the second period (2009–present), EV policies were legitimized through climate goals and aimed to create a market for EVs, regardless of where these EVs originated from (Ryghaug and Skjølsvold 2019). Today, the first phase tends to receive little appraisal, but in our view, this phase was crucial if we want to understand how Norway became the most advanced laboratory for EVs that it currently is if you compare the market share of electric vehicles to other countries across the world.

The first modern visions of Norway as a country producing electric vehicles emerged in the 1970s as a response to the OPEC oil embargo of 1973. Norway had vast hydropower resources. Developing an EV industry and challenging the fossil fuel-based mobility was therefore seen by a set of industrial pioneers as an ideal way of strengthening energy security in Norway. The notion was that while Norway at the time was poor in oil, it was rich in hydropower, and this should be reflected in the country's preferred mode of transport (Asphjell, Asphjell, and Kvisle 2013).

While these ideas did not come to fruition, it illustrates how an interest in developing new technology was shaped by a combination of international developments and local concerns. The activities also established a network of industrial actors who sustained the vision of Norway as an EV producing nation. This vision re-gained momentum in the early 1990s primarily as a response to the enactment of the Zero Emission Vehicle (ZEV) legislation in California, which established a credit-system where car dealers had to earn credits from the sale of non-emission vehicles to legally be able to continue selling petrol cars (Hoogma et al. 2002). The scheme has later been dubbed "one of the most daring and controversial air quality policies ever adopted" (Collantes and Sperling 2008). Transition scholars have highlighted that the legislation was geared towards 'innovation pull', producing 'windows of opportunity' for battery electric vehicles (Kemp 2005), and has been deemed central to the development of EV friendly policies in other countries, such as Japan (Åhman 2006).

In Norway, the allure of this new potential market resulted in the emergence of a set of new industrial alliances, where Norwegian manufacturers, the hydropower industry and a series of other actors were able to mobilize national and international funding to produce a small, plastic chassis urban EV—a 'personal independent vehicle' called PIV (Hoogma et al. 2002; Buland 1994; Andersen 2013). For us, this illustrates that the shaping of pilots can be highly local, but that actors, interests and policies might also affect pilot activities at a vast distance. In turn, the opposite might also be true: through establishing a relatively large pilot market, the Norwegian context has arguably influenced the innovation work done by large international auto manufacturers.

During the early 1990s, the development and testing of these vehicles in Norway and the US took the form of several technology pilot projects, resembling those described earlier in this chapter. Building on a combination of financial resources from the EU, Norwegian industry and public funds, the goals of these pilots were, on the one hand, to verify technology and learn from real-life conditions, while, on the other hand, serving as a tool for public engagement. In many ways, these pilots can be interpreted as relatively successful, in the sense that they demonstrated the potential for EVs under cold conditions in Norway, as well as under quite hot and urbanized conditions in California. Just as important for the emergence of Norway as the EV-laboratory we see today, however, was the fact that these processes built substantial visions and expectations for a future large scale electromobility industry based in Norway, which was in the interest of several actors, including the hydropower industry (Skjølsvold and Ryghaug 2020). As a result, the Norwegian authorities throughout the 1990s gradually introduced policy incentives that were, on the one hand, meant to stimulate this industry, but which on the other hand, were meant to stimulate demand for the vehicles that this emerging industry was producing.

These developments led to the emergence of a pioneer niche market in Norway, in which vehicles such as the CityBee (later rebranded as Th!nk) and the Kewet found its niche role on Norwegian urban roads. These vehicles never became mainstream despite efforts to ramp up policies to stimulate their demand. Towards the mid-2000s, the Norwegian EV industry actors had more or less given up, and the Norwegian EV market was mainly served by international actors who considered the Norwegian EV benefits lucrative. This, however, was clearly not the end of the Norwegian EV laboratory. Rather, by this time, the climate issue had gained much higher prominence than it had during the 1990s, which changed the discourse in Norway substantially.

Since Norway was largely powered by renewables due to its vast hydropower energy supply, policy makers quickly turned their eyes towards transport in their quest to decarbonize, and electro mobility soon became one of the main strategies of decarbonization. Towards the end of the 2000s industry leaders like Mitsubishi, Peugeot, Citroën and Nissan began launching new flagship EV models and Norwegian car dealers immediately began importing them. The Norwegian EV market especially boomed after the introduction of Mitsubishi i-MiEV in 2010 and the Nissan LEAF in 2011 (Lorentzen et al. 2017).

Since 2009, Norway has actively embraced and been outspoken about the country's role as an EV policy laboratory, in which new policies locally and nationally have been experimented heavily with. This has been a combination of policies intended to stimulate increased demand for EVs, locally and nationally, and policies to enable an easy transition from driving fossil fuelled cars to electric cars. This package of policies has included free or reduced cost on ferries and VAT exemption for car leasing. Further, a governmental support scheme for public charging infrastructure was implemented in the years 2009–2010, followed by public coordination of fast charging infrastructure and charging facility developments across the country. Small municipalities with few chargers can today seek financial support, and the goal is to have fast charging stations around every 50 km on Norwegian roads. The network of chargers throughout the country is probably a culturally important safety-net which mitigate what is commonly referred to as range anxiety (Noel et al. 2019) and is something that contribute to the further growing of the EV market. Several municipalities and cities have also followed Oslo in allowing EVs to drive in bus lanes.

Through the activities discussed above, Norway has effectively become a sort of large-scale pilot project that explores the societal consequences of implementing electromobility earlier than comparable countries. As the above discussion have alluded to, this laboratory role was shaped by a long and cumbersome process, which includes some distinctly local interests such as those rooted in hydropower and national industrial development, international research and development networks and consortia, as well as, later, the pressure to reduce climate emissions. As a national pilot, the goals are also much more diverse, than for example the goals of a pilot in a physical laboratory. In a national project, one tests technology in a real-life setting. In addition, one also tests the effects of policies, behaviours and practices at large scale, as well as the links between developments in, until now, quite separately working sectors such as energy and transport. An important example is experimenting on the way that electromobility is and might impact the electricity grid, the operation of the grid, as well as innovations in managing and developing the electricity grid (see Chap. 3 for an example).

The Significance of How Demonstration Projects Are Shaped

In the above, we have produced a set of accounts that illustrates how pilot and demonstration projects are shaped by societal processes. We have worked from assumptions derived from decades of STS-scholarship, which highlights how historical, social and cultural dimensions as well as the interests of involved actors shape technologies (e.g. Pinch and Bijker 1984). Technologies, in turn, shape and produce future societal conditions, indicating that technology and society co-produce each other (Jasanoff 2004). To us, this suggests that technologies take on an often-unappreciated political significance in shaping future social order (Winner 1980), and that how, by whom and where technologies are made is essential for the shaping of these future social orders.

The discussions above suggest that pilots are important within energy transition activities, and within the shaping of future societies. Therefore, it is central to understand who the actors involved are, which agendas they advance and legitimate and the processes through which such advancement and legitimation occurs. While there are significant differences between the projects we have discussed, there are also a series of similarities, which to us points towards the importance of thinking systematically about how to shape and produce pilot and demonstration projects differently in the future.

All the projects discussed here starts from the assumption that technologies are the key vehicles through which societies will reach their energy and climate ambitions. The projects reflect on the relationship between their innovation activities and social aspects to various degrees, but they all mobilize a rather narrow understanding of the potential ways their activities and technologies relate to the social world. Insofar, we have also seen the social world represented primarily as technology users or as supporters of unfolding innovation activities. To us, this is reflective of the actors and processes that shape the types of pilot and demonstration projects that we have probed here. In this way, the pilot projects can be said to serve as a mirror, projecting back the interests of the involved actors in the form of a technologically oriented image of societal change to the world around.

While the projects are technology oriented, they mobilize these technologies to address different types of issues. Some of these issues, such as the need to balance supply and demand in energy systems are generic and related to dynamics of the energy transition. Others, such as demographic transitions and related challenges in healthcare are local, more specific and points towards the potential of broader social engagement within and around pilot projects. This dual focus is reflective of the resources and actors that tend to be involved. On the one hand, projects tend to involve international funding, large national and international actors within energy, ICT and research, but on the other hand, often also local interests and locally invested actors. In such constellations, we see the mobilization of local issues as hopeful, because it points towards the possibility of broader social and political transformative repertoires enacted through technology (see e.g. Marres 2016; Ryghaug et al. 2018). In the projects discussed above, however, this potential largely remains dormant. While the established pilot projects reflect the local conditions and involved interests, the imagination with respect to the role of the social and political is limited, and mainly operationalized as technology use or consumption. This signals a narrow agenda within such pilots, and to us opens a question if the types of constellations we have discussed above will be able to deliver radical transformations or mainly advance agendas of incremental change.

Seen all together, our discussions on "the origin stories" about how different pilot projects came into being and how they were produced, sheds light not only on discrete innovation practices, but illustrate a key function of pilots beyond testing and deployment of new technologies. In this chapter, we have noted how pilots serve to "mirror" the interests that produce them. This metaphor, however, is too weak because pilots not only reflect, but also amplify the interests, resources and rationalities that are built into them. As a social and material performance, pilot projects signal what Mike Michael (2000) has described as the temporal proximity of futures rooted in these pilots. In other words, pilot projects signal that the world will soon be changing in their image.

This points to the importance of re-thinking how pilot and demonstration projects are currently produced. Given the need to radically transform our collective relationships to energy and transport, we believe it is worth questioning if the types of projects that we discuss here ask the right questions, or if one could envision innovation done differently. Schot and Steinmueller (2018) question the contemporary quest for technological innovation as a solution to the sustainability challenge, and Marres (2016) notes that piloting or experimenting for sustainability can be done with roots in social questions, as opposed to those primarily rooted in technology. We have also flagged hopefulness on behalf of the political potential of pilot projects earlier in this chapter. On the one hand, the responsibility for change lies in the hands of project operators. On the other hand, our discussion also suggests that there is a broader question to be raised here, concerning research and innovation policy. The origin stories discussed above do not appear in an epistemic vacuum but follows broader logics of how change and transition is promoted in contemporary research and innovation efforts in Europe. We will return to the issue of innovation policy in Chap. 4.

Beyond Pilots: Understanding Pilot Projects in Broader Energy and Sustainability Transitions

So far, we have been concerned with the construction of pilot and demonstration projects. Now, we will turn to what such projects do once they have been established, and the potential relationships between pilot projects and the world around. Pilots situate new technologies in society (Forlano 2019), providing a potential socio-technical model for how to organize activities in ways that are considered more sustainable or more beneficial than contemporary practice. However, their reach is often limited. Above we have discussed examples rooted in the laboratory, a few houses in a neighbourhood, islands and a nation. A key question for innovators is how such limited pilot activities can be expanded or scaled up to transform broader elements of society (see e.g. Naber et al. 2017; Ryghaug et al. 2019). In our case, these questions first translate into an interest in the processes and work that makes such projects grow, and second, into a concluding discussion on the potential politics that are advanced as such projects become increasingly important for the direction of societal development.

Upscaling and Accelerating Energy Transitions Through Pilot Projects?

In current debates on sustainability transitions, pilot projects and experiments have been pointed to as central drivers in the acceleration of transitions (Von Wirth et al. 2017; see Chap. 1 for a discussion on the acceleration phase of transitions). The question of how pilot projects and experiments can 'scale up' has therefore been highlighted as a key research challenge for scholars working on sustainability transitions over the coming years (Köhler et al. 2019).

Within STS, processes that bear resemblance to what transition scholars call up-scaling have been conducted for decades. The most famous example is perhaps Latour's (1993) account of how Louis Pasteur transformed his laboratory trials on Anthrax vaccines into a successful program for the whole of France, thus stopping regular mass death of livestock in the country. In Latour's interpretation, this work consisted of Pasteur conducting a series of stagings, in which farmers were enrolled and required to re-produce laboratory-like conditions in their farms through strict measures of hygiene and cleanliness. Following this, trials were conducted in-situ on farms, both as a real-life test, and as demonstrations of the vaccine's feasibility. This means that up-scaling, in this case, entailed expanding the laboratory into new realms of society through partnering with farmers, and convincing these farmers to adopt the practices of the laboratory. Beyond this, Pasteur also enrolled the mass media, which on different occasions produced vivid accounts of the vaccination success. For Latour, then, the successful upscaling of Pasteur's laboratory experiment primarily hinges on the production of a public (in this case the farmers) who shares the interests of the innovator, who in Latour's narrative becomes just as much a social and political entrepreneur as a microbiologist.

The process where innovators and entrepreneurs seeks to attract interest and support for a particular innovation by means of persuasion, negotiation and aligning with interests of other actors, is often described as translation (Callon 1984; Latour 1987). Re-read in light of sustainability transition challenges, this model of innovation is a way to describe processes of up-scaling and acceleration of transitions. Translation describes the processes of how certain actors might ascribe new roles to other actors, and highlights that the route towards technological success lies in building strong actor-networks around new artefacts. Building the network consist of developing different scenarios and enrolling new actors in the enactment of such scenarios (Latour 1987). When a scenario is developed, the scenario is translated to appeal to what is believed to be the relevant actors' needs and wishes. Translation has been described as a four stage process that emphasizes the displacements and transformations of goals, interests, and devices, human beings and inscriptions happening in four 'moments' or phases with four components: problematization, interessement, enrolment and mobilization (Callon 1984; Hess 1997).

Problematization signifies the process of defining the issue in a way so that other actors accept one's definition of the problem. They gradually come to accept one's knowledge claims or technology as an obligatory point of passage, that is, as a necessary means to solve their problem. Interessement refers to imposing and stabilising the roles of the other actors defined by one's problematization. In other words, the process of translating the images and concerns from one world into that of another, and then disciplining or maintaining that translation in order to stabilise a powerful network, thus "locking" other actors into the roles that were proposed for them in the actor's programme for resolving that problem. The result of interessement is the enrolment where actors or entities are attached to the network in interrelated roles. Finally, mobilization is ensuring that supposed spokespersons for relevant collective entities are properly representative of all members of the network that are acting as a single agent, representatives to act as spokesperson of other entities (Callon 1984). From Latour's work, enrolment does not only mean involving people, but also nature and technologies.

Within sustainability transition studies, the last years have seen several examples of scholars exploring similar processes around pilot and demonstration projects dealing with new energy and mobility technologies. Here, however, the focus on translation tends to be implicit, while the literature rather works to develop typologies which point to ideal-typical patterns of development, or of mechanisms, that result in the growth of projects.

Naber et al. (2017) provides one example, as they distinguish between four patterns of upscaling. Growing, describes a continued process of experimentation, through which more actors become participants, or market demand increases. This might enable growth either in the size of projects, or in the types of activities involved. Replication points to the use of key concepts from one trial in other locations. Accumulation indicates the production of links between pilots and networks of pilots. Finally, transformation points to changes that occur as such projects shape wider institutional configurations.

Frantzeskaki et al. (2017) provide a similar line of reasoning and develop a framework for understanding the different ways that urban transition initiatives can accelerate transitions. This framework describes five mechanisms that may contribute to acceleration. These five mechanisms are: Upscaling, which entails growing the number of members, supporters or users of a single transition initiative; Replicating, which is the take up of new ways of doing, organizing and thinking of one transition initiative by another transition initiative; Partnering, which is the pooling and/or complementing of resources, competences, and capacities in order to exploit synergies between initiatives; Instrumentalising, which entails tapping into and capitalizing on opportunities provided by the multi-level governance context of the city-region; and finally, Embedding, which describes the alignment of old and new ways of doing, organizing and thinking in order to integrate transition initiatives into city-regional governance patterns.

We find these accounts useful, because they give some hints about what up-scaling might entail in practice. However, we also question whether the neutrality of terms like "patterns" and "mechanisms" really captures the dynamics of up-scaling. Our discussion on the genesis of pilot and demonstration projects suggests that the projects reflect and amplify the interests of those who produce them. STS scholars such as Bruno Latour (1993) and Michael Callon (1984) have pointed to the social and political aspects of spreading ideas and technologies, which points to a much more active role on behalf of certain actors in advancing their ideas, and that an element of this is also persuading others who might contest the technology at hand (e.g. Sørensen et al. 2018). The focus on the active elements, for example, in the form of translation when looking at the up-scaling of pilot and demonstration projects also makes the political character of this work more explicit: ultimately, up-scaling entails actively working to promote one vision of future socio-technical order at the expense of others. In what follows, we will look at one example of how such a process might look in practice.

How Pilots Scale Up: An Example

Pilot projects often remain standalone learning sites where little knowledge travels beyond the project and the involved participants. This means that many projects bring little change to the broader systems they intend to transform (Heiskanen et al. 2017). Naber et al. (2017) and Frantzeskaki et al. (2017) provide useful starting points for contemplating how movement beyond individual trials might, and do, unfold. From the early days of STS (e.g. Callon 1984; Latour 1987), we are also sensitized to the political and social entrepreneurship that might enable such processes in action. The example we now zoom in on is a pilot project that is primarily pushed forward and conducted by one company. This company is one of the largest grocery wholesalers in Norway. The company sells and distributes groceries to 1700 stores and restaurants, a task it has conducted using traditionally fuel intensive delivery trucks.

In contemporary news media coverage and public discourse, this company clearly stands out as the main character in a narrative about an ambitious pilot project that combines large-scale production of solar power with electrolysis to produce hydrogen fuel cells in a quest to decarbonize its large fleet of heavy trucks. Their role as such, however, was not always clear-cut, because the initial phases of this pilot were not set-up by the company. Rather, the company was arguably the main public for a trial envisioned by other actors, primarily a group of technology developers and scientists who had been researching and advocating the benefits of the hydrogen economy for more than a decade.

This group had developed a small fuel cell, which was intended to power only the lifts that distribution trucks use when loading and unloading goods. As heavy users of such trucks with a reputation for seeking out low-carbon technology, the company constituted what the scientists imagined as ideal users of this fuel cell. The problem was that the company was not really interested, as they thought hydrogen was the fuel of the future, but not the present. Further, they were not convinced by the merits of replacing the lift battery with an alternative. Hence, to stretch the activities of the scientists beyond the laboratory, the company needed convincing. This illustrates that these scientists recognized the strength of a pilot project. Hydrogen has always been the fuel of the future in Norway (Kårstein 2008), but with few pilot projects to learn from. A pilot could materialize the vision (Engels and Münch 2015), positioning the hydrogen economy temporally closer to our present time (see. e.g. Michael 2000).

Faced with this challenge, the scientists attempted another strategy, namely by proposing a small project to measure emissions from urban delivery trucks. The company accepted to be part of this project, and the results indicated that on a typical urban delivery route, trucks spent more time idling with the engine running than driving around, because the diesel engine was used to power the electrical lift battery. With these results in hand, the grocery wholesaler became an interested actor, who envisioned both economic and environmental benefits from replacing the batteries with a small hydrogen fuel cells to power the lift used when loading and unloading goods of their distribution trucks. They agreed to retrofit a few trucks. This can be interpreted as an initial form of upscaling in the form of what Naber et al. (2017) calls growing. The process of achieving this growth echoes Latours (1993) analysis of Pasteur, where "interessement" and "enrolment" is achieved through extending the laboratory to new sites in society.

Technically, the initial trials were a success, and the company was able to document substantial emissions reductions, due to the reduced need for idling. Just as importantly, this small-scale pilot transformed the management of the grocery wholesaler to hydrogen enthusiasts, in a way reminiscent of Callon's (1984) notions of translation. The company was now not only using the technology from the initial trial but were actively producing new visions for the hydrogen economy in Norway. A step towards this goal would be their own pathway towards a fully hydrogen powered fleet of trucks. Hence, this meagre technology trial yielded lessons that were now transforming the foundations of this company. Frantzeskaki et al. (2017) calls this embedding, while with the words of Naber et al. (2017) this was arguably transformative, in the sense that this initial pilot contributed to transforming what Geels and Schot (2007) have described as the grammar, or the rules of the game for this company.

These developments unfolded at the same time as the company was conducting a much simpler technology project, namely investing in a large solar power park which was to be placed at the rooftop of one of the grocery wholesalers storage facilities. In certain periods of the day and season, this setup produced a surplus of electricity on their site. Coupled with visions about hydrogen futures, the idea of using this renewable energy source to produce hydrogen through electrolysis soon emerged through discussions with the earlier mentioned scientists and technology developers.

Following this, the goal was to establish a new pilot project which aimed to transform the way this grocery wholesaler transported goods. The vision was that the vehicles would be entirely fuelled by hydrogen produced on-site. This is symptomatic of a form of institutional change that has been identified in the German energy transition, where increased interaction between the renewable energy sector, the transportation sector and ICT actors have created what Canzler et al. (2017) have described as a new strategic action field which opens for new and sometimes radical forms of innovation. A challenge, however, was that acquiring heavy duty hydrogen trucks was difficult: as far as the involved actors knew, such trucks did not exist.

The company and the scientists set out to partner with a large European truck manufacturer but received only lukewarm interest. The car manufacturing industry has been described as a conservative regime that has slowly and reluctantly reoriented towards implementing new and more sustainable fuels and technologies (e.g. Penna and Geels 2015). While the grocery wholesaler is a relatively large company in Norwegian terms, it was a small international player, and struggled to find anyone willing to take part in what was perceived as a high-risk endeavour. Through intense lobbying together with the scientists, however, the company was able to convince their existing car manufacturer, to deliver three 27-tonne trucks, to be experimentally developed by the supplier together with a project group consisting of people from the grocery wholesaler and associated scientists. This move can be said to represent a new form of growth within this project, perhaps best described in terms of what Frantzeskaki et al. (2017) call partnering, or in Naber et al.'s (2017) terms, accumulation, signalling that there were now even more elements drawn into this conglomerate of activities.

Following this, the company added several new elements to the pilot. They invested in an off-the-shelf hydrogen production facility which they installed on-site to produce hydrogen from solar power, as well as a fleet of hydrogen powered forklifts to be used within their own storage facility. These activities arguably resulted in what Naber et al. (2017) call transformation: The grocery wholesalers' role in the energy and transport systems was changed drastically. From being a traditional wholesaler that stores and delivers groceries around the region, they were now a large electricity producer, a producer of hydrogen fuel cells, and a co-developer of a new type of delivery truck. Thus, from the initial small beginnings within a laboratory and shifting the engine on the lifts of a few trucks, this is clearly a story of up-scaling. Still, however, the technologies and innovation processes of this single pilot only encompass a few dedicated actors. But the efforts to upscale go beyond the story of this company.

Currently, the grocery wholesaler arguably pursues three key strategies to produce better framework conditions for the innovations they have been part of developing and testing. The first consists of continued growth within their own organization, primarily through expanding the fleet of hydrogen powered trucks. In doing this, they also model future hydrogen demand, to illustrate to others that they will be unable to meet their own demand after 2023. This brings us to the second strategy, which can be characterized as a form of translative networking or partnering, where they attempt to persuade actors within renewable energy production such as wind farms to produce hydrogen. Finally, the company's third strategy consists of politicizing their own pilot activities and their own engagement for hydrogen as an essential element in the future socio-technical order of

Norway. They lobby municipalities in order to increase the importance of environmental aspects in procurement processes. Further, they regularly seek to influence annual state budgets, policies and authorities responsible for providing the framework conditions and supporting activities to cut emissions and promote more sustainable transportation. They actively target political parties and national strategy processes, working to convince others that hydrogen should have priority over competing technology options.

It is difficult to assess the success of all the activities described above, but this is also beside our point. Instead, the activities illustrate how the work and dynamics of constant growth and upscaling, within and around a pilot project can unfold. All this work has been highly visible in parts of the Norwegian public debate. It has attracted significant attention from political actors, and the Norwegian prime minister has been photographed in front of this grocery wholesaler's hydrogen trucks, proclaiming that this project heralds no less than the start of a new era. From meagre beginnings where engineers were seeking out application areas for testing and experimenting with a small hydrogen fuel cell and with the goal of powering a small vehicle as a trial, the actors involved in this pilot project now seek to transform their own practical, technological and institutional surroundings in such a way that these surroundings in turn might become precisely what enables the project and this version of the hydrogen economy to grow further.

Implications for the Literature on Upscaling: From Patterns and Mechanisms to Strategies

Within the socio-technical sustainability transitions literature, there has been a tendency to describe processes of up-scaling in relatively neutral terms such as patterns (Naber et al. 2017) or mechanisms (Frantzeskaki et al. 2017). This is a tempting move, because it provides hopes of generalizability, which in turn opens for learning across cases. We sympathize with this ambition, but working from an STS perspective, we also want to problematize this. The case discussed above illustrates that this apparent neutrality might conceal the political character of enacted agency within processes up-scaling. Our case indicates that in all observed instances, upscaling was challenging, and that other outcomes were also possible. In the spirit of deriving lessons from case studies, we therefore want to point to two strategies that we see as essential in work to scale up from pilot and demonstration projects.

- Persuasion: a key strategy to achieve several of Naber et al.'s (2017) and Frantzeskaki et al.'s (2017) patterns and mechanisms is persuasion. This signals that pilot and demonstration projects are potentially contestable spaces. They represent a potential socio-technical order, but there are many such orders out there, including the dominant ones that pilots often seek to destabilize. Thus, partnering, growth, replication and accumulation are not only patterns or mechanisms, or necessarily the result of peaceful alignment and deliberation, but outcomes of successful acts of persuasion.
- Politicization: a key element of Naber et al.'s (2017) framework is the pattern of transformation, whereby a pilot project transforms wider institutional conditions. Politicizing is an explicit way of problematizing and challenging current institutional configurations based on ongoing innovation activities. It signals broad social and political ambitions beyond technology deployment, and explicitly targets actors and processes that are outside the unfolding pilot activity as such.

These two points are suggestive of the potential of mobilizing STS in both the study of sustainability transitions, and as inspiration for practical strategies within such transitions. While perspectives rooted in the MLP and related frameworks tends to highlight the need to produce shared visions and directionality, our narrative here also point towards the importance of agonism and contestation, and that in light of this, translation becomes a key mode of working.

Concluding Discussion: Towards an Appreciation of the Political Character of Pilots and Demonstration Projects

The discussions in this chapter have illustrated how pilot and demonstration projects are shaped and produced, as well as how actors work to scale-up such pilots. We have argued that pilot projects embody and enact desires for future socio-technical order. Through these discussions, we have come to see how pilots tend to reflect, legitimate and amplify the interests and resources that shape them. Further, we have seen that processes of scaling up such projects are characterized by persuasion and politicization. Through this, we have come to see pilots as sites where future possibilities become materialized, where transitions and transformations are enacted.

With this as a backdrop pilot projects within energy and mobility must be considered political sites, or as Marres (2016) calls them, in-between sites, where normal obligations and relations may cease to exist. Within such in-between sites one can, according to Marres, explore (a) technologies, through implementing unproven concepts, (b) politics, by suspending established rules of public accountability and (c) society, by trying to establish new ways of doing things (Ibid., p. 4). Seen through such a lens, some types of pilot projects might take the form of a materialized imaginary (Engels and Münch 2015), in the sense that they not only render visible a potential technological future, but also a desirable future society around and intertwined with the technology.

This creates a form of tension. On the one hand, it might result in pilot projects being very effective tools of governance and innovation (Marres 2016). They can be productive in terms of introducing new technologies and new socio-technical trajectories and futures. The examples we have discussed in this chapter to a large degree confirms this. However, it is less clear if these pilot and demonstration projects are effective in terms of fulfilling other societal and social needs and how this mode of governance through pilot projects might affect how we steer societies more broadly. For us, these observations lead to an interest in a set of aspects that characterize pilot projects as sites of governance or as sites of societal steering.

First, the discussions in this chapter have illustrated that developing and conducting pilot and demonstration projects tends to re-distribute the legitimacy of actors within a specific domain. One obvious example of this is the increasing importance of ICT competence in transforming domains such as energy and mobility.

This reflects the way pilot projects often combine a desire for highly specialized technological outcomes with broad and complex social change. Thus, pilot projects need combinations of very specific competence just to work technologically, while they often need other actors to provide the societal vision within and around which these technological solutions are intended to work. Hence, pilots not only provide legitimacy to new sets of actors, but also to new sets of visions, ideas and goals for future society. As we have seen in this chapter, the content of such visions might differ radically: Visions and expectations might range from visions of a convenient and simpler everyday life for the elderly, to more prosperous island-life and multi-and cross-sectoral visions for the future hydrogen society. Across such visions, there tend to be new roles for technology users, customers or citizens, new forms of business or new modes of organizations, which in sum serve to re-shape many aspects of society within or around pilot projects.

This means that the potential political effects of pilot projects stretch far beyond the boundaries of such projects and allows actors that innovate within energy and transport to assume the power to transform aspects of society that at a first glance might appear out of their reach. This is one of the reasons why it is essential to probe participation within and around emerging technologies and such projects, and also why we dedicate an entire chapter in this book to this theme where we ask if pilot projects represent a shift away from democratic decision-making in society, and if so—how should we understand this development? How should the development be governed?

Frantzeskaki et al. (2017) note that pilot projects and other transition initiatives are key vehicles to accelerate sustainability transitions, in other words, to increase the speed of technology deployment, diffusion and thereby the replacement of existing and fossil-intensive socio-technical systems. According to the sociologist Hartmut Rosa (2013) acceleration is one of the key traits of late modern societies. In his analysis, acceleration is a generic processual trait that has fundamentally altered the way political decision making has been conducted in modernity. In a time that values acceleration beyond other imperatives, he argues, democratic decision making is too slow, and serves as a "brake" to further acceleration.

Rather than democratic decision making, he highlights that decisions with respect to collective ethics and future organization for realizing such ethics are privatized, that key societal decisions are increasingly made based on economic evaluations and that key disputes are settled through judicial processes rather than democratic deliberation. This amounts to a description of a time characterized by a post-political situation. Some readings of governance rooted in pilot and demonstration projects have highlighted that such projects can be understood to represent precisely this: That such projects "intentionally sidestep the tensions between bottom-up and top-down approaches to innovation in favour of lateral partnerships" (Evans and Karvonen 2011), and that they therefore provide ample opportunities to remove control from government, to cede it to private interests, and to understand climate change primarily as a business opportunity "in the guise of helping society at large" (Evans and Karvonen 2011).

Such a sinister reading of the potential political role of pilot projects, however, is countered by accounts that highlight how they might come to occupy more of an intermediary space, where they allow for new types of negotiations and deliberations between technology developers and potentially implicated actors. Here, pilot projects might serve to open technology driven processes to wider and more diverse forms of democratic participation, rooted in diverse and often local issues. This means that innovation through such processes might also serve to challenge established structures, and though this, bring about new configurations which would otherwise not emerge. Evans and Karvonen (2011) have concluded that the politics of such projects are "up for grabs". This is a reading we sympathize with. In our view the politics of such projects are always up for grabs, because what they will be, depends on the specific configurations that shape the process by which they are established, in which resources are mobilized, as well as by the way such projects are positioned vis-à-vis broader societal processes. In the next chapter, we will follow up on this by more explicitly probing pilots as potential sites of participation.

REFERENCES

Åhman, M. (2006). Government policy and the development of electric vehicles in Japan. *Energy Policy 34*(4), 433–443.

Andersen, O. (2013). Towards the Use of Electric Cars. In *Unintended Consequences of Renewable Energy* (pp. 71–80). London: Springer.

Asphjell, A., Asphjell, Ø., & Kvisle, H. (2013). *Elbil på Norsk*. Oslo: Transnova.

Ballo, I. F. (2015). Imagining energy futures: Sociotechnical imaginaries of the future Smart Grid in Norway. *Energy Research & Social Science, 9*, 9–20.

Barnett, J., Burningham, K., Walker, G., & Cass, N. (2012). Imagined publics and engagement around renewable energy technologies in the UK. *Public Understanding of Science, 21*(1), 36–50.

Berker, T., & Throndsen, W. (2017). Planning story lines in smart grid road maps (2010–2014): Three types of maps for coordinated time travel. *Journal of Environmental Policy & Planning, 19*(2), 214–228.

Broto, V. C., & Bulkeley, H. (2013). Maintaining climate change experiments: Urban political ecology and the everyday reconfiguration of urban infrastructure. *International Journal of Urban and Regional Research, 37*(6), 1934–1948.

Buland, T. (1994). Framtiden er elektrisk. *IFIM-notat, 4*, 94.

Bulkeley, H. A., Broto, V. C., & Edwards, G. A. (2014). *An urban politics of climate change: Experimentation and the governing of socio-technical transitions.* Routledge.

Callon, M. (1984). Some elements of a sociology of translation: Domestication of the scallops and the fishermen of St Brieuc Bay. *The Sociological Review, 32*(1_suppl), 196–233.

Canzler, W., Engels, F., Rogge, J. C., Simon, D., & Wentland, A. (2017). From "living lab" to strategic action field: Bringing together energy, mobility, and information technology in Germany. *Energy Research & Social Science, 27,* 25–35.

Cetina, K. K. (1995). Laboratory studies: The cultural approach to the study of science. In *Handbook of science and technology studies* (pp. 140–167). Los Angeles: Sage Publishing.

Christensen, T. H., Ascarza, A., & Throndsen, W. (2013). Country-specific factors for the development of household smart grid solutions: Comparison of the electricity systems, energy policies and smart grid R&D and demonstration projects in Spain, Norway and Denmark. IHSMAG Project report. Retrieved May 28, 2020, from https://vbn.aau.dk/ws/files/168269618/Christensen_et_al._Country_specific_factors_2013.pdf.

Collantes, G., & Sperling, D. (2008). The origin of California's zero emission vehicle mandate. *Transportation Research Part A: Policy and Practice, 42*(10), 1302–1313.

Cotton, M., & Devine-Wright, P. (2012). Making electricity networks "visible": Industry actor representations of "publics" and public engagement in infrastructure planning. *Public Understanding of Science, 21*(1), 17–35.

Curtis, M., Torriti, J., & Smith, S. T. (2018). Demand side flexibility and responsiveness: Moving demand in time through technology. In *Demanding energy* (pp. 283–312). Cham: Palgrave Macmillan.

Engels, F., & Münch, A. V. (2015). The micro smart grid as a materialised imaginary within the German energy transition. *Energy Research & Social Science, 9,* 35–42.

ETIP SNET. (2018). Vision 2050: Integrating smart networks for the energy transition: Serving Society and Protecting the Environment. Retrieved March 22, 2020, from https://www.etip-snet.eu/wp-content/uploads/2018/06/VISION2050-DIGITALupdated.pdf

European Commission (2020) *Horizon 2020.* Work Programme. Brussels. Accessed from: https://ec.europa.eu/research/participants/data/ref/h2020/wp/2018-2020/main/h2020-wp1820-intro_en.pdf

Evans, J., & Karvonen, A. (2011). Living laboratories for sustainability: Exploring the politics and epistemology of urban transition. In *Cities and low carbon transitions* (pp. 126–141). London: Routledge.

Forlano, L. (2019). Cars and contemporary communications| Stabilizing/destabilizing the driverless city: Speculative futures and autonomous vehicles. *International Journal of Communication, 13*, 28.
Fosso, O. B., Molinas, M., Sand, K., & Coldevin, G. H. (2014). Moving towards the smart grid: The Norwegian case. In *2014 International Power Electronics Conference (IPEC-Hiroshima 2014-ECCE ASIA)* (pp. 1861–1867). IEEE.
Frantzeskaki, N., Borgström, S., Gorissen, L., Egermann, M., & Ehnert, F. (2017). Nature-based solutions accelerating urban sustainability transitions in cities: Lessons from Dresden, Genk and Stockholm cities. In *Nature-based solutions to climate change adaptation in urban areas* (pp. 65–88). Cham: Springer.
Frøysnes, A. S. (2014). *Bare en jævla boks?: en analyse av visjonsarbeidet knyttet til avanserte måle- og styringssystemer (AMS)* (Master's thesis, NTNU)
Gangale, F., Vasiljevska, J., Covrig, C. F., Mengolini, A., & Fulli, G. (2017). *Smart grid projects outlook 2017*. The Netherlands, Petten: Joint Research Centre of the European Commission.
Geels, F. W. (2002). Technological transitions as evolutionary reconfiguration processes: A multi-level perspective and a case-study. *Research Policy, 31*(8–9), 1257–1274.
Geels, F. W., & Schot, J. (2007). Typology of sociotechnical transition pathways. *Research Policy, 36*(3), 399–417.
Goulden, M., Bedwell, B., Rennick-Egglestone, S., Rodden, T., & Spence, A. (2014). Smart grids, smart users? The role of the user in demand side management. *Energy Research & Social Science, 2*, 21–29.
Grydehøj, A., & Kelman, I. (2017). The eco-island trap: Climate change mitigation and conspicuous sustainability. *Area, 49*(1), 106–113.
Haugland, B. T. (forthcoming) Self driving imaginaries, politics and innovation. Under review in Palgrave Communications.
Heidenreich, S. (2015). Sublime technology and object of fear: Offshore wind scientists assessing publics. *Environment and Planning A, 47*(5), 1047–1062.
Heiskanen, E., Hyvönen, K., Laakso, S., Laitila, P., Matschoss, K., & Mikkonen, I. (2017). Adoption and use of low-carbon technologies: Lessons from 100 Finnish pilot studies, field experiments and demonstrations. *Sustainability, 9*(5), 847.
Hess, D. J. (1997). *Science studies. An advanced introduction*. New York: New York University Press.
Hoogma, R. J., Kemp, R., Shot, J., & Truffer, B. (2002). *Experimenting for sustainable transport. The approach of strategic niche management*. Spon Press: London and New York.
Jasanoff, S. (Ed.). (2004). *States of knowledge: The co-production of science and the social order*. Routledge.
Kårstein, A. (2008). *HyNor–den norskehydrogenveien?: En studie av en storteknopolitisk hybrid* (PhD Thesis, NTNU).

Katzeff, C., & Wangel, J. (2015). Social practices, households, and design in the smart grid. In *ICT innovations for sustainability* (pp. 351–365). Cham: Springer.

Kemp, R. (2005). *Zero emission vehicle mandate in California: Misguided policy or example of enlightened leadership* (pp. 169–191). UK, Cheltenham: Edward Elgar.

Köhler, J., Geels, F. W., Kern, F., Markard, J., Onsongo, E., Wieczorek, A., ... Fünfschilling, L. (2019). An agenda for sustainability transitions research: State of the art and future directions. *Environmental Innovation and Societal Transitions, 31*, 1–32.

Kohler, R. E. (2002). Place and practice in field biology. *History of Science, 40*(2), 189–210.

Latour, B. (1987). *Science in action: How to follow scientists and engineers through society*. Harvard University Press.

Latour, B. (1993). *The pasteurization of France*. Harvard University Press.

Latour, B., & Woolgar, S. (1979). *Laboratory life: The construction of scientific facts*. Princeton University Press.

Lorentzen, E., Haugneland, P., Bu, C., & Hauge, E. (2017, October). Charging infrastructure experiences in Norway-the worlds most advanced EV market. In EVS30 Symposium (pp. 9–11)

Marres, N. (2013). Why political ontology must be experimentalized: On eco-show homes as devices of participation. *Social Studies of Science, 43*(3), 417–443.

Marres, N. (2016). *Material participation: Technology, the environment and everyday publics*. Springer.

Michael, M. (2000). Futures of the present. In N. Brown, B. Rappert, & A. Webster (Eds.), *Contested futures* (A sociology of prospective technoscience) (pp. 21–39). Aldershot: Ashgate.

Morozov, E., & Bria, F. (2018). *Rethinking the smart city*. New York: Rosa Luxemburg Stiftung.

Naber, R., Raven, R., Kouw, M., & Dassen, T. (2017). Scaling up sustainable energy innovations. *Energy Policy, 110*, 342–354.

Nahuis, R., & Van Lente, H. (2008). Where are the politics? Perspectives on democracy and technology. *Science, Technology, & Human Values, 33*(5), 559–581.

Noel, L., de Rubens, G. Z., Sovacool, B. K., & Kester, J. (2019). Fear and loathing of electric vehicles: The reactionary rhetoric of range anxiety. *Energy Research & Social Science, 48*, 96–107.

Papazu, I. (2016). Authoring participation. *Nordic Journal of Science and Technology Studies, 4*(1), 17–31.

Papazu, I. (2018). Storifying Samsø's renewable energy transition. *Science As Culture, 27*(2), 198–220.

Park, S. (2011). Iceland's hydrogen energy policy development (1998–2007) from a sociotechnical experiment viewpoint. *International Journal of Hydrogen Energy, 36*(17), 10443–10454.

Penna, C. C., & Geels, F. W. (2015). Climate change and the slow reorientation of the American car industry (1979–2012): An application and extension of the Dialectic Issue LifeCycle (DILC) model. *Research Policy, 44*(5), 1029–1048.

Pinch, T. J., & Bijker, W. E. (1984). The social construction of facts and artefacts: Or how the sociology of science and the sociology of technology might benefit each other. *Social Studies of Science, 14*(3), 399–441.

Rosa, H. (2013). *Social acceleration: A new theory of modernity.* Columbia University Press.

Rosenow, J., & Kern, F. (2017). EU energy innovation policy: The curious case of energy efficiency. In *Research handbook on EU energy law and policy*. Edward Elgar Publishing.

Rosol, M., Béal, V., & Mössner, S. (2017). Greenest cities? The (post-) politics of new urban environmental regimes. *Environment and Planning A: Economy and Space, 49*(8), 1710–1718.

Ryghaug, M., & Sørensen, K. H. (2009). How energy efficiency fails in the building industry. *Energy Policy, 37*(3), 984–991.

Ryghaug, M., & Toftaker, M. (2016). Creating transitions to electric road transport in Norway: The role of user imaginaries. *Energy Research & Social Science, 17,* 119–126.

Ryghaug, M., & Skjølsvold, T. M. (2019). Nurturing a regime shift toward electro-mobility in Norway. In *The Governance of Smart Transportation Systems* (pp. 147–165). Cham: Springer.

Ryghaug, M., Skjølsvold, T. M., & Heidenreich, S. (2018). Creating energy citizenship through material participation. *Social Studies of Science, 48*(2), 283–303.

Ryghaug, M., Ornetzeder, M., Skjølsvold, T. M., & Throndsen, W. (2019). The role of experiments and demonstration projects in efforts of upscaling: an analysis of two projects attempting to reconfigure production and consumption in energy and mobility. *Sustainability, 11*(20), 5771.

Sadowski, J., & Levenda, A. M. (2020). The anti-politics of smart energy regimes. *Political Geography, 81,* 102202.

Schot, J., & Steinmueller, W. E. (2018). Three frames for innovation policy: R&D, systems of innovation and transformative change. *Research Policy, 47*(9), 1554–1567.

Silvast, A., Williams, R., Hyysalo, S., Rommetveit, K., & Raab, C. (2018). Who 'uses' smart grids? The evolving nature of user representations in layered infrastructures. *Sustainability, 10*(10), 3738.

Skjølsvold, T. M. (2014). Back to the futures: Retrospecting the prospects of smart grid technology. *Futures, 63,* 26–36.

Skjølsvold, T. M., & Ryghaug, M. (2015). Embedding smart energy technology in built environments: A comparative study of four smart grid demonstration projects. *Indoor and Built Environment, 24*(7), 878–890.

Skjølsvold, T. M., & Ryghaug, M. (2020). Temporal echoes and cross-geography policy effects: Multiple levels of transition governance and the electric vehicle breakthrough. *Environmental Innovation and Societal Transitions, 35*, 232–240.

Skjølsvold, T. M., Ryghaug, M., & Berker, T. (2015). A traveler's guide to smart grids and the social sciences. *Energy Research & Social Science, 9*, 1–8.

Skjølsvold, T. M., Ryghaug, M., & Throndsen, W. (2020). European island imaginaries: Examining the actors, innovations, and renewable energy transitions of 8 islands. *Energy Research & Social Science, 65*, 101491.

Solbu, G. (2018). The physiology of imagined publics. *Science & Technology Studies, 31*, 39–54.

Sørensen, K. H. (2004). Cultural politics of technology: combining critical and constructive interventions?. *Science, technology, & human values, 29*(2), 184–190.

Sørensen, K. H. (2013). Beyond innovation. Towards an extended framework for analysing technology policy. *Nordic Journal of Science and Technology Studies, 1*(1), 12–23.

Sørensen, K. H., Lagesen, V. A., & Hojem, T. S. M. (2018). Articulations of mundane transition work among consulting engineers. *Environmental Innovation and Societal Transitions, 28*, 70–78.

Strengers, Y. (2013). *Smart energy technologies in everyday life: Smart Utopia?*. Cham: Springer.

Strengers, Y. (2014). Smart energy in everyday life: are you designing for resource man?. *Interactions, 21*(4), 24–31.

Throndsen, W. (2017). What do experts talk about when they talk about users? Expectations and imagined users in the smart grid. *Energy Efficiency, 10*(2), 283–297.

Tøndel, G., & Seibt, D. (2019). Governing the elderly body: Technocare policy and industrial promises of freedom. In *Digitalization in industry* (pp. 233–259). Cham: Palgrave Macmillan.

Vesnic-Alujevic, L., Breitegger, M., & Pereira, A. G. (2016). What smart grids tell about innovation narratives in the European Union: Hopes, imaginaries and policy. *Energy Research & Social Science, 12*, 16–26.

Von Wirth, T., Fuenfschilling, L., Frantzeskaki, N., & Coenen, L. (2019). Impacts of urban living labs on sustainability transitions: Mechanisms and strategies for systemic change through experimentation. *European Planning Studies, 27*(2), 229–257.

Williams, R., & Edge, D. (1996). The social shaping of technology. *Research Policy, 25*(6), 865–899.

Winner, L. (1980). Do artifacts have politics? *Daedalus, 109*, 121–136.

Open Access This chapter is licensed under the terms of the Creative Commons Attribution 4.0 International License (http://creativecommons.org/licenses/by/4.0/), which permits use, sharing, adaptation, distribution and reproduction in any medium or format, as long as you give appropriate credit to the original author(s) and the source, provide a link to the Creative Commons licence and indicate if changes were made.

The images or other third party material in this chapter are included in the chapter's Creative Commons licence, unless indicated otherwise in a credit line to the material. If material is not included in the chapter's Creative Commons licence and your intended use is not permitted by statutory regulation or exceeds the permitted use, you will need to obtain permission directly from the copyright holder.

CHAPTER 3

Democratic and Participatory Pilot Projects?

Abstract This chapter starts from the normative assumption that since pilot projects are key sites in the shaping of future societies, it is essential that they are conducted in an inclusive and democratic way. Building on key perspectives from STS, we focus on two aspects: First, we consider participation as an orchestrated and distributed phenomenon, highlighting the fact that the way actors participate in such innovation activities will be shaped by technologies, assumptions and the work of a series of actors related to pilot projects. Consequently, we also note how new forms of participation can be actively nurtured. Second, we explore the role of technologies in shaping material participation. Here, we explore how material traits might produce new forms of awareness, knowledge or literacy, and new practices or action, amounting to what we call *energy citizenship*.

Keywords Pilot projects • Participation • Orchestration • Energy citizenship

In the last chapter we made the case that pilot, and demonstration projects are political entities. On the one hand they may be sites that seemingly promote relatively narrow technological agendas, but on the other hand they also formulate and materialize future socio-technical orders across an increasing span of societal sectors. Through our discussion we illustrated how pilot projects legitimate and amplify the interests and resources that produce them. We indicated that many pilot projects tend to be centred around a few technology oriented interests, and that formulations and enactments of social relations within such projects are often limited to consumption or technology use.

It is with this as a backdrop that we now turn our attention to pilot and demonstration projects as potential sites of participation. On our behalf this is both a normative and an analytical move. It is analytical in the sense that it allows us to probe another facet of such projects, normative because we follow Delina and Sovacool (2018) and other scholars who highlight that transitions should not only entail implementing new technologies and phasing out old ones, but that this should be done in a just way. Achieving just transitions requires the mobilization of a plurality of voices in processes of innovation.

While sustainability transition scholars working from the multi-level and related frameworks (e.g. Geels 2005, 2010b; Kemp et al. 1998) have primarily been preoccupied with the process where technologies are developed from a narrow and alternative niche to an established and powerful regime (see Chap. 1 for a more detailed discussion), STS has a long tradition of asking precisely how technoscientific processes can be opened to the participation of broader publics. The interest in this comes from observations that scientific expertise and technology developers often overestimate the universality of their proposals (e.g. Wynne 1996; Irwin and Wynne 2003; Ryghaug et al. 2011), and that this might result in the production of technologies that do not work in practice, and scientific facts that misses out on insights anchored in everyday lives and other rationalities than the technoscientific. A relevant example here could be the development of a technology that seeks to transform the way energy is consumed in households, without mobilizing insights about the lives within households in the production of knowledge and the production of technology.

Michel Callon (1999) has sketched a development where ideas about public participation in scientific and technological development processes have been expanded to encompass three models for public participation. The first and classical model is the public education model. Here, publics do not participate in the production of technology or scientific facts. Rather, publics are informed through acts of education, with the rationale that if they only knew the benefits of new developments, they would accept them. Hence, this is a model that builds on technocratic ideals and signals that public knowledge deficiencies are barriers to progress. The second model described by Callon is the public debate model. Here, technologies and scientific proposals are publicly debated, often through mechanisms such as public hearings, consensus conferences of citizen panels. This is a form of democratization which opens decision making to public scrutiny, and which assumes that the voices of different interests can enrich and improve decision making. Callon first two models have in common that they render public participation as a process that is as external to the actual production of knowledge or technology (Chilvers and Kearnes 2015; Marres 2012).

Finally, Callon (1999) points to the most radical model, which is the co-production model. Here, publics are not only allowed to debate technoscientific developments that are of concern to them at a distance, they are actively mobilized in the process of technoscientific development and production. This latter model has become significant in what some scholars have dubbed 'the participatory turn' in technoscience (e.g. Felt and Fochler 2010), where narratives of increased and new forms of public participation is expected to improve the relationship between technoscience and society.

Over the last years, such accounts of public participation in technology processes have been expanded further, in part because of criticisms that have emerged against the models discussed above. Approaches such as participation in the form of public debate or processes of co-production have been pointed to as institutionalized to the point of becoming political machines (Barry 2001), that produces publics through "offering" fixed discursive and practical spaces of engagement (Felt and Fochler 2010). Hence, in this book we take inspiration from scholars who zoom out from a focus on discrete sites where participation is enacted towards studying participation as a *distributed phenomenon* which is produced by a range of diverse actors across sites (Chilvers and Longhurst 2016; Chilvers and Kearnes 2015; Chilvers et al. 2018). A key element in these perspectives that are of crucial importance to us is the roles that technologies and materiality play in articulating interests and publics, and through this also enabling new forms of participatory practices (Marres 2016).

These insights lead us to a dual focus in the remainder of this chapter. First, we will probe the ways that participation is produced or *orchestrated* by a range of actors (e.g. Skjølsvold et al. 2018). This is a move that illustrates not only that participation is a distributed phenomenon, but which points to the distributed responsibility of making conditions that cater for a diversity of voices, rationalities and practices of participation. Second, we will look at how participation might be *enacted*. Here, we give significance to the enactment of a form of material participation that we discuss as *energy citizenship* (Ryghaug et al. 2018), a term intended to open participation in energy transition activities to broader and more politically informed actions.

THE ORCHESTRATION OF PARTICIPATION IN PILOT AND DEMONSTRATION PROJECTS

As discussed in Chap. 2, many pilot and demonstration projects reflects the involved actors strong focus on technology development and deployment, with a limited focus on potentially transformative effects such as the ability to scale up, or what Marres highlights as the potential to actively tinker with political and societal aspects. Many projects only marginally focus on 'real' technology users, which means, that insights from users do not feed into further technology development. One consequence of this that has been observed in the research literature is that prospective "publics" or "users" are cast as groups that strongly resemble the technology developers (Strengers 2014; Skjølsvold and Lindkvist 2015). In instances when publics are imagined differently, they tend to be envisaged as barriers to success, with the preferred scenario often being automated technological solutions that work in the background without being noticed (see also Fjellså et al. forthcoming). Such projects, it seems, rejects all models of participation proposed by Callon (1999).

Despite of such observations, much of the *rhetoric* surrounding pilot and demonstration projects tends to be anchored in notions of involvement, active engagement and user centric design. These are signifiers that all point towards high levels of participation. Funding bodies increasingly also demand that technology developers take measures to include knowledge from technology users in new projects. This is reflected more broadly in European energy policy, which highlights that a key goal of the energy transition is to make future energy systems 'citizen centric' (Ingeborgrud et al. 2020). Yet, citizenship promoted through such rhetoric tends to be reduced to finding new ways of making people act as rational agents in economic markets (e.g. Wallsten and Galis 2019). How can we begin an analytical and normative process of working towards alternative models of participation?

Our account is inspired by recent scholarship within STS, which highlights that participation is co-constructed, relational, emergent and in the making (Chilvers and Kearnes 2015). Within such an understanding, participation is not the individual act of opting into or opting out of a particular technology trial, but an outcome of a process involving a wide array of actors and objects (see also Marres 2016). Participation, then, emerges in interaction between actors, in-situ, which amplifies a long-standing point made by STS-scholars that participation is not external to technoscientific endeavours, but rather an integral aspect constituted by scientific practice (e.g. Shapin and Schaffer 1985). A metaphor that makes this point explicit is that of *orchestration*. Orchestration points to how the work of actors who conduct pilot and demonstration projects seeks to produce specific types of participation. In sum, our interest here is in "participation in the making", and especially the ways that participation-making entail attempts at producing new forms of social and political order.

This interest leads us to ask what the consequences of such a move would be. On the one hand, participation within such a framework is a phenomenon that is subject to the same processes of shaping and construction that we discussed for technologies in Chap. 2. This means that the way participation looks in a project is contingent on the cultural, historical, social, economic and technological specificities of the project, as well as the work done by actors in that particular site. This does not mean that participation is a phenomenon fully constituted by local aspects. It is also shaped by the distributed work of actors—actors working from a long distance. Examples of this include the work of national and international policy makers, large companies and international organizations. Thus, within such a perspective, participation is both localized and specific, and connected to and embedded within wider circuits of actors and networks across scales (Chilvers and Kearnes 2015). An important consequence of this line of reasoning is that we come to see participation as a collective endeavour constituted by collectives of participation, and wider ecologies of participation.

Pilot and demonstration projects are a particularly suitable case in point here for exploring participation in the making. Orchestration within this context consists of two processes. First, enrolment, which "refers to the way in which different (human and non-human) actors are drawn into a particular form of participatory collective practice and definition of the issue at stake" (Chilvers and Longhurst 2016, p. 591). The second is mediation, which refers to "the way in which a participatory collective is held together by different devices, processes, skills, or 'technologies of participation'" (ibid., p. 591). In what follows, we will explore how these processes might emerge within and around pilot projects that seek to advance technologies associated with energy and sustainability transitions.

If we briefly re-visit the pilot projects discussed in Chap. 2, we can describe them in terms of processes of enrolment and mediation. The pilot projects are shaped and constructed by a series of actors and resources, who formulate quite different issues that the pilot projects are expected to address. Our example of a smart grid pilot shaped in part by local healthcare interests can serve as an example. Here, actors within the energy and ICT sector first enrolled healthcare actors in the work to produce issues to address though smart grid piloting. This participation was mediated through a series of workshops, which sustained the healthcare workers participation in the production and stabilization of this issue. In turn, this constellation made necessary the enrolment of prospective technology users, who would participate primarily by using the technology. Here, this participation was mediated through the pilot technologies, but also through a broader political interest both in how to deal with the local demographic transition, and in how to make electricity consumption smarter. Hence, orchestration is multidirectional, multiple and enacted by a range of actors.

In the past we have identified collectives of actors engaged in such work as collectives of orchestration (Skjølsvold et al. 2018). Such collectives are primarily identified through the activities that they engage in. On the one hand, the activities of these collectives are often anchored in very localized spaces or institutional settings, where they participate in the transition through engagement with concrete issues. Many of the pilot projects discussed in Chap. 2 contain a series of such issues: how the electricity grid handles the influx of renewables, making technology that caters for the needs of the healthcare system, working to produce a new piece of hardware to be tested in a laboratory, or working to produce a new standard or piece of legislation. In doing so, these collectives, on the one hand, perform acts of participation. They participate in transition-oriented activities through their work. On the other hand, they also engage in targeting and aiming to transform practices beyond their own immediate site and situation, working to format the participation of other actors at a spatial and temporal distance. Thus, we come to see participation as co-produced, but also as a politicized phenomenon. What participation is, what it should be and what the goals of participation should be can be contested and transformed.

Arguably, there are a set of at least three ideal typical collectives of orchestration (Skjølsvold et al. 2018), that work to orchestrate participation at a spatial and temporal distance around the types of projects that we discuss in this book. These three are (a) collectives of policy production and regulation, (b) collectives of research, development and innovation, (c) and collectives of technology design. In the following, we give some illustrative examples of how such collectives work to orchestrate participation within pilot and demonstration projects, primarily within the energy domain.

Collectives of Policy Production and Regulation

In Chap. 2 we discussed how policies and regulations are resources for and enablers of innovation activities. One way of elaborating on this, is to highlight how policies seek to orchestrate participation in energy transition activities across time and space. An example of these dynamics is found within the field of smart energy technology.

Through the production of policies and regulations, actors have for a long time worked to format how ordinary citizens engage with energy. Through policies of liberalization and privatization policy makers have tried to cultivate active and rational economic consumers (Karlstrøm 2012; Silvast 2017). Policies and regulations stimulating the implementation of smart energy technologies can be seen as an expansion of these logics, aiming to produce consumers who act flexibly and rationally as economic agents, e.g. by buying and selling electricity, as well as changing the timing of their consumption and thus providing flexibility to the energy system (Ballo 2015; Skjølsvold 2014; Christensen et al. 2020). Hence, policies, for example for smart meter implementation produce visions of future societies where the use of smart meters is widespread and produce visions that render the desired effects of such use visible. Through this, such policies seek to enable a specific form of participation in future markets on behalf of customers.

Hence, one can say that the production of these policies seeks to orchestrate how citizens engage with energy. However, and just as importantly, policy production is also central in orchestrating the work of actors within the energy industry, ICT industry and others who work to establish smart energy pilots. In Chap. 2 we saw an example of this from the island of Samsøe, where local authorities formulated local policies and visions with the aim of attracting external innovators. Similarly, the policy priorities of the European Commission and other funding bodies are of vital importance in shaping the strategies of such innovators. Beyond such priorities, innovators and technology implementers take cue from policy and regulation, particularly in questions regarding the implementation of technological standards. In sum, this points to collectives of policy production and regulation as key in enrolling a range of various actors in participating in energy transition-oriented pilot activities in different ways.

Collectives of Research, Development and Innovation

Pilot projects and demonstration projects have been central for the development of smart energy technologies over the last years. Through these developments such projects have arguably been instrumental in orchestrating the participation of ordinary households and citizens, particularly through developing and testing technologies, price schemes and modes of organization that seek to change timing or character of energy consumption often referred to as demand-response or demand side management. In such projects, ordinary households can typically participate in transition activities as consumers, for example by responding to price signals. However, smart energy pilot projects practically do this in quite different ways. Some projects explicitly seek very active engagement on behalf of citizens, who are expected to work as energy managers within their own homes. In other instances, pilots orchestrate a more passive form of participation, where new tasks are delegated, for example to new automatic technologies. Hence, pilot projects are a way of enrolling citizens in participation, often through using new technologies or engaging in new types of consumption of prosumption.

The actors who establish pilots, however, do not necessarily only seek to orchestrate the way technology users participate in transitions. As we saw in Chap. 2, the actors behind such projects might also become more politically ambitious, and seek to transform practices of policy making and public procurement. In such instances we saw how innovators worked to produce visions of future societies anchored in their own technologies, and thereby worked to enrol policy makers and legislators in co-producing a reality that renders those very technologies part of a more plausible alternative future. Thus, the actors involved have ambitions of transforming the participation of technology users, policy makers, and in turn, also other actors within their industry.

Collectives of Design

In the discussion above, the orchestration of participation by policy makers and R&D actors unfold through the production of networks of material devices, organizations and visions. Collectives of design form around similar concerns, but are anchored more concretely in the production of the specific things that make up for example pilot projects. Hence, collectives of design produce interfaces, switches, apps, screens and other concrete objects. The issue around which they form relates to how these technological objects should be shaped to do effective work in the kinds of networks discussed above.

In STS, design has for long been understood as a form of orchestration at a distance, through the notion of scripting (Akrich 1992) which more recently has also been linked explicitly to the production of public participation (Marres and Lezaun 2011). The orchestration of participation through design is arguably a two-step process of (1) producing visions of future technology use (Borup et al. 2006), and (2) translating the consequences of such visions into concrete objects. Within smart energy pilot projects such as those discussed in Chap. 2, one can distil at least four ideal-typical user-characters that designers often mobilize in the production of technologies. The first character can be described as 'greedy', a trait translated into participation through a rational form of consumption. This entails visual devices focusing on numbers (money saved, money earned, kilowatt hours not used), and graphs communicated via in-home displays, apps, bills or websites, providing information on consumption and production levels as well as costs and income. The second character is understood as politically motivated and green, driven by a desire to mitigate environmental problems and participate politically as a citizen, a mode of participation increasingly identified as promising, and that we will discuss at length later in this chapter (e.g. Devine-Wright 2012; Ryghaug et al. 2018). These users are often enrolled through scripts that provide information about CO_2 savings, communicated through apps, displays and websites, thus providing what could be understood as means to enact *energy citizenship* through material participation (Ryghaug et al. 2018).

The third user character is 'simple'; disinterested in technology or energy use, with main motivation related to comfort and convenience.

Such users are often enrolled through scripts that minimize the need for active input, for example as pre-programmed household settings like 'night', 'day' or 'home' and 'away'. Participation is often delegated to technology, under the assumption that their disinterest would be a threat to the goals of the system (see also Chilvers et al. 2018). Finally, some projects envisage a 'social' or collective user, imagined operating in groups (multi-person households, neighbourhoods, communities). This character is orchestrated to participate primarily through two mechanisms: competition or cooperation. Scripts targeting this group often enrol users by focusing on political engagement and citizenry concerning environmental issues. These issues can be presented on online platforms that consist of gaming elements or competitions (gamification), or discussion forums, where platform users would be encouraged to discuss openly their experiences with solar energy, energy savings and environmental issues in general (Skjølsvold et al. 2018).

Technology Users as Orchestrators of Participation

In discussions about the implementation of new technologies, or in discussions about energy transitions more broadly, the users of technology are often discussed in terms of social acceptance (e.g. Wolsink 2018; Ingeborgrud et al. 2020). Discussions about acceptance can provide important insights. As an example, models of participation that see citizens primarily as consumers requires citizens to accept and act on new price signals, often with new technologies to work. However, technology users or citizens can take on other roles than receivers of ready-made solutions. Arguably, they can also advance innovation, and orchestrate the participation of other actors in the energy transition. In past studies (Throndsen and Ryghaug 2015; Throndsen et al. 2017) we have argued that as households become enrolled in pilot projects, they might re-interpret the purpose of such a project, thereby also transforming the future direction of work within the project. An example of this can be that a technology-oriented project becomes re-defined as a project that is also a political endeavour, in the sense of communicating the virtues of sustainability.

Another example can be found in the discussions of Norway as a national laboratory for electric vehicles discussed in Chap. 2. Here, early technology users arguably played a vital role in shaping the national policy landscape. They did so by organizing in new ways and explicitly targeting policy makers and industry in attempts to make them re-think the way private transportation was conducted in Norway (Ryghaug and Skjølsvold 2019; Skjølsvold and Ryghaug 2020). Further, these technology users, through their use of and communication about EVs worked to influence the desires and expectations of other Norwegian drivers. Through this work, they were also part of establishing a critical mass of enthusiasts that enabled a new EV market, which in turn made EV production more attractive for international automobile producers.

Implications of Perspectives from Studies of Ecologies of Participation, Contestation and Orchestration

Through a focus on policy production, research and development and design over the last paragraphs, we have come to see participation as a phenomenon beyond individual choice and individual technology encounters, but rather as carefully orchestrated activities, distributed across an ecology of participatory collectives. Further, we have seen how the technology users have an active role in shaping their own modes of participation, and that they are important for shaping wider spaces of participation, including a potential role as co-orchestrators of the work in other collectives. This analysis has both theoretical and practical implications. Theoretically, the discussion serves to raise some challenges to the multi-level perspective (MLP) (Geels 2002) as the ways participation unfolds through the staging and carrying out of innovation activities, as discussed above, are not easily categorized within a niche-regime scheme. If one interprets the MLP strictly, one would expect transitions in the organization of participation to, above all, grow out of protected or nurtured niches, or through new practices formed in alternative energy communities gradually breaking into and destabilizing regime-level norms and behaviours. However, in the analysis we have conducted, we find that instances of participation are in-fact partially produced and shaped through work of actors that in accordance to the MLP would be considered regime actors (e.g. policy makers, large construction companies, DSOs). Thus, our focus on co-production, orchestration and situated analysis provide a more heterogeneous narrative, where the potential of nurturing desired traits such as participation can be done by a broad range of actors across different levels or domains in society.

The practical implications of this are substantial, but also surprisingly simple. On one level, this discussion highlights the merits of working to link different forms of practice, across collectives, epistemic foundations and through different technologies and objects in innovation endeavours. Our analysis, as the analysis of other STS scholars, suggest that this is currently done too conservatively and that it is too uniformly rooted in dominant understandings of participation (Chilvers et al. 2018). In other words, we find very strong networks of policies, institutions, research programs and new technologies supporting participation as individual consumption and behaviour change, while the collectives around more experimental and radical forms of participation are fewer.

Accordingly, we should not think that simply linking existing policy-making, R&D, design and households would serve to produce radical new forms of participation, or that this would result in participatory practices rooted in new concerns. An ecological understanding of participation opens up a broader understanding of what experimentation in this domain could entail. Such an understanding stresses that experimentation may be seen as something beyond testing new technologies in and around households. Thus, experimentation to increase participation in energy transitions might also entail trying new ways of producing policy, standards and regulations, new modes of working within R&D, and experimental design practices. In other words, we believe there is an active role to be played here for innovation policy. We will return to this point in Chap. 4.

An Object-oriented Perspective: Material Participation

A potential critique of our focus on participation as a distributed phenomenon is that through re-casting the activities of many actors as participation, we risk losing sight of involved power dynamics. It might be argued against our case that a symmetrical perspective where the activities of citizens, large corporations and policy makers are all understood as different forms of participation, might lead to the conclusion that one should not prioritize any of these groups in the pursuit of democratic legitimacy. Rather than reducing the importance of citizens, our aim here is to elevate their position. The shift from pure individual choice to distributed responsibilities helps us achieve this, as it actually opens for a broader repertoire of action.

An emphasis on technology users has been highlighted in various streams of literature that focus on concepts such as public acceptance or user acceptance of new technologies, public perception of new technologies, and public engagement with new technologies. Such literatures implicitly point to technology users as a barrier to the success of technologies (Karlstrøm and Ryghaug 2014; Heidenreich 2015). As a contrast, the concept of energy citizenship has been put forward (Devine-Wright 2012) to highlight the potential political engagement of citizens in the face of environmental and climatic issues. We believe there is much to be gained by linking the concept of energy citizenship to developments in STS that provide an object-oriented account of the constitution of politics and publics (e.g. Latour 2005; Marres 2016; Throndsen and Ryghaug 2015). In such accounts, publics and issues materialize and are constituted around things that might enable new forms of material participation. Building on this, we propose this as a promising avenue for exploring material energy citizenship (see also Ryghaug et al. 2018).

Energy Citizenship as Means of Material Participation

The concept of energy citizenship exemplifies the growing strand of research, arguing that energy transitions require active citizen participation and not only passive acceptance (Ingeborgrud et al. 2020). Adding to this, we bring the concept of material participation from STS into this discussion (Marres 2016). Material participation is an 'object-oriented' or 'device-centered' perspective that focuses on the role of technologies and material objects for participation in political matters of concern. The concept of material participation has grown from a body of STS inspired work on 'Dingpolitik' (Latour 2005), highlighting the ways material objects enable the configuration of issues, concerns and publics, thereby potentially producing new ways of representing diverse interests and voices around such concerns. This literature considers publics and issues to be emergent rather than static and highlights the political potential of technologies as objects that might both enact a political reality and intervene in the world (Marres and Lezaun 2011) for example by "enrolling actors such as local communities, governmental organizations and environmental researchers" (Marres 2013, p. 427). Material participation, then, can be thought of as a specific mode of engagement (Marres 2012). Thus, by linking the concepts of material participation and energy citizenship we have wanted to turn the discussion to the ways new technologies can offer new means of enacting concerns when it comes to issues like climate change and sustainability.

More concretely, material energy citizenship emerges as objects enable the formation of a set of concrete capacities and competences that are in part shaped by the new material realities and technologies. These capabilities are:

- The formation of new forms of awareness
- The formation of new forms of literacy or knowledge
- The formation of new practices and actions

These capabilities allow the formation and enactment of political projects in everyday life (Ryghaug et al. 2018). However, as Martiskainen et al. (2018) has pointed out, there is no determinism in this: one should not confuse the existence of material participation with general assumptions about the social and political potential of certain objects. Instead, technological objects can acquire a range of political capabilities as they become part of different relations, and different configurations. This also suggests that technologies can become mobilized actively in the orchestration work of actors such as policy makers and researchers. To us, this is a hopeful notion because we believe recognizing that objects might have political significance might open new avenues of design and innovation. In what follows, we will look briefly at one example, namely how electric vehicles might enable the formation of a new form of energy citizenship.

The Co-production of Energy Citizenship in Collectives with Electric Vehicles

At a first glance, electric vehicles might resemble an electric duplicate of its fossil counterpart. It has a similar form and function: it is a car that takes its driver and passengers from one destination to another. Yet, electric vehicles often take on more significance for its owners. For many, the electric vehicle serves to formulate a more active political and practical engagement in relation to energy and climate issues (Ryghaug and Toftaker 2014). Electric vehicles might enable new forms of sensitization or awareness with respect to one's mobility, new types of literacy or knowledge with respect to energy and climate issues and finally the formation of new practices and new actions. In this way, electric vehicles can facilitate a new form of energy citizenship (Ryghaug et al. 2019).

The constitution of such capabilities is related to the material qualities of electric vehicles, which can be seen at different stages of using the car. Electric vehicles are powered by a battery that needs charging, and the driving range of most EV models is limited compared to petrol cars. In the process of buying an EV, this means that many are immediately confronted with their current driving habits, and the need to reconcile these habits with what EVs allow for. After having acquired the vehicle, many EV drivers need to incorporate new attentive practices, such as monitoring battery levels and planning when and where to charge, in order to make the vehicle function in everyday life (Ryghaug and Toftaker 2014; Ingeborgrud and Ryghaug 2019). Trips must also be planned according to battery status and the availability of charging infrastructure. Thus, once purchased, the EV, with its batteries and charging infrastructure, enters into collectives of mundane mobility, materializing the issues of energy scarcity and mobility needs, typically constituted through visual displays in the car that indicate battery levels and remaining driver distance or mobile phone apps providing similar information. At a practical level, EV drivers might become sensitized to minute-by-minute electricity use while driving. In turn, this allows for self-evaluation of the efficiency of driving styles and opens for experimenting with new driving styles to increase energy efficiency (Anfinsen et al. 2019).

These material qualities of EVs can serve to problematize current mobility habits and for some open up for evaluating understandings of normal mobility patterns and driving ranges, as well as more active engagements with mobility alternatives such as biking, walking and public transport (Ryghaug and Skjølsvold 2019). In some instances, the introduction of EVs and the actualization of availability of electricity and infrastructure as important issues have amplified an interest in participating in other energy transition activities such as improving home energy efficiency or local energy production. Some EV drivers' for instance, take an interest in acquiring in-home battery technologies or local micro-production of electricity such as solar panels (Ingeborgrud and Ryghaug 2019).

In Chap. 2, we discussed how new institutional links between the renewable energy industry, the transport industry and the ICT industry have opened a new strategic action field, where new modes of innovation might unfold (Canzler et al. 2017). Our discussion here illustrates how links between renewable energy and transportation can also provide impulses for change in the everyday life of ordinary households. Hence, what we see here might resemble a mundane form of sector-coupling, pointing towards the importance of new modes of action amongst such ordinary householders.

In sum, EVs can represent a material and discursive actualization of climate and environmental issues. As such, acquiring and driving EVs constitute a tangible way to act upon the climate change issue for some, through more environmentally oriented consumption. However, electric vehicles also take on other roles, which are important for the formation of energy citizenship. Electric vehicles frequently become contested, and their sustainability has been subject to substantial public controversy (Ortar and Ryghaug 2019). For many, these controversies become yet another way to participate in transition activities. Many EV owners report that they are often forced to defend their transportation choices, hence engaging in everyday deliberation over how a future transportation system should look, in ways they would never have to if they drove a petrol vehicle. Some report "reading up" as a deliberate strategy of planning for such encounters. Armed with new knowledge, some become active proponents of environmental arguments in their local communities, promoting electric mobility, but also other ways of participating in energy transitions.

Above, electric vehicles have been discussed as a technology that may evoke material participation and energy citizenship. Other objects and technologies such as solar panels and in-home displays and feedback systems have also proven to have similar qualities in certain contexts (see Ryghaug et al. 2019 for more on this). However, there is no determinism in such technologies producing participation. Further, participation does not automatically lead to justice. While an electric vehicle can be constitutive of energy citizenship, electric vehicles are also, for many, expensive, and hence provide a means to participate for those that are already well-off. A challenge, therefore, is to mobilize the dynamics of material participation not only around luxury items for the few, or in wealthy countries, but to actively work to produce a material reality that caters for broad and socially transformative material participation.

Summing up, energy citizenship consists of a set of concrete capacities and competences: the formation of new forms of awareness, the formation of new forms of literacy or knowledge, and the formation of new practices and actions. This avenue of research illustrates the benefits of not thinking of publics only as pre-existing entities formed around discrete or pre-defined issues. Instead, things, devices and technological objects can constitute issues such as those related to climate change and environmental issues. They can produce new ways of engaging with the world, and hence there is a need for broader scholarly engagement with the role of objects in producing citizenship. This is important also for actors who orchestrate participation through new devices such as smart energy technologies. These actors tend to frame and orchestrate participation as consumption. Material participation and energy citizenship suggests that a much broader repertoire is possible and that we should focus more on object issue-oriented publics, who come into the political arena to take part in constructing scientific and technological futures (Marres 2007; Stilgoe et al. 2014; Ryghaug et al. 2019).

Pilot Projects and the Production of Collective Energy Citizenship

The above discussion and the literature on material participation and energy citizenship have tended to focus on the potential of mundane objects that internalize and enable action primarily within and around individual households. Examples are electric vehicles, solar panels and smart energy meters (Throndsen and Ryghaug 2015; Ryghaug et al. 2019), eco homes and mobile phone chargers (Marres 2016). Our focus on pilot and demonstration projects allows us to make expansions to arguments that have been made within this arguably device or 'gadget centred' literature.

We will now expand on the concept of material participation and energy citizenship, by looking at an example of how the socio-technical configurations of pilots might enable new forms of citizenship. In doing so, we are amongst other things able to discuss how the capacities and qualities of specific places not only serve as resources for pilot project innovators, but as reservoirs of potential political issues around which engagement might form. This indicates, on the one hand, that the material, cultural and practical specificities of places might be generative for innovation, and that the same capacities can serve as basis for sketching out explicitly political technoscientific endeavours. Hence, energy citizenship is potentially a more collective phenomenon which does not only depend on the relationship between technology user and technology/gadget, but which is also shaped by the broader material and institutional setting of pilot and demonstration sites. Hence, we expand on insights which highlight the link between systemic innovation and resonance with political realities of everyday life and mundane experience.

The gradual shift from centralized and fossil-based production sites to more decentralized and distributed systems based on renewables will likely make electricity production a mundane matter for increasing numbers of people, as discussed in Chaps. 1 and 2. This may create new types of interaction between traditional energy suppliers and citizens, producing new roles and actor constellations throughout the system. This type of decentralization will typically include new modes of renewable energy production (microgeneration), micro grids, local storage solutions, automation, feedback technologies (such as energy displays) and combinations of such technologies (Parag and Sovacool 2016; Skjølsvold et al. 2018; Ryghaug et al. 2019). Can these dynamics feed into the formation of new types of energy citizenship and material participation?

The Material Political Dynamics of Shared of EV-charging

One example of how this development can unfold in practice can be found through following the Norwegian EV-transition. On the one hand, this development entails implementing electric vehicles. We have discussed this development in Chap. 2, as well as in the last section of this chapter. Here, we are interested in aspects of this transition beyond the immediate situation of driving and buying a car, and rather zoom in on what is arguably the emergence of a new form of systemic material form of participation and politics which forms around the charging of EVs. This exemplifies that introducing EVs at large scale represents a systemic change affecting different sectors as well as affecting the institutional organization of social life.

Here, we are particularly interested in the relationship between EV implementation and changes in the electricity grid. Electrification of transport entails increased electricity demand. Historically such challenges have been addressed by expanding the electricity grid and the electricity production capacity. Today it is has become just as common to address this issue by working to transform the character of the electricity demand, typically thorough smart charging technologies aimed at providing more flexible timing of EV-charging. In the Norwegian case, the dynamics of implementing flexibility are arguably creating a new form of localized politics around the implementation and use of such charging infrastructures. This is particularly visible in areas that are organized as a form of condo or collective housing, which is a quite common form of home ownership in Norway and the other Nordic countries. In such arrangements people own the apartments, while buildings and other infrastructure such as parking, garages and other outdoor spaces are typically owned communally, often managed by an elected board of housing representatives.

Over the last years, such housing condos have seen a massive influx of EVs. This indicates that EVs are in the Norwegian case no longer a niche for the wealthiest. The EV-surge, therefore, provides a means of participating in the energy transition, and serves as a potential enabler of mass individual energy citizenship. On the other hand, the influx of EVs have also enabled a more collective form of energy engagement in such areas, because implementing charging infrastructure is subject to collective forms of decision-making. Garages and parking spaces are often a scarce resource and part of the shared community space in condos and apartment buildings; this means that expenses related to these infrastructure arrangements are typically also shared among dwellers. Similarly, local electricity grids are often pressed for capacity, which means that many neighbourhoods are forced to take an active stance towards local electricity grid management. This means the influx of EVs also transforms the role of actors that in the past have not been engaged or interested in energy issues—inhabitants and elected boards now become central in making decisions about the implementation of new, complex energy infrastructures.

On one level, this issue may be seen as a quite straightforward infrastructural capacity problem on how to ensure that enough power capacity is available to charge all EVs that are connecting to electricity grids. Smart charging has been pointed out as one solution to these capacity challenges, and is currently being implemented by pioneering neighbourhoods, that for our purposes serve as, what we in Chap. 2 discussed as geographically bound pilots. In what follows we will discuss the emergence of one such pilot, by enquiring about the resources mobilized and processes of legitimation, as we did in Chap. 2. Here, however, we emphasize the emergence of a new network of material and discursive elements, and the way that this pilot re-configures the local material-politics of energy transition participation.

The Role of National Legislation in the Orchestrating New Material Political Dynamics

While we emphasize the material aspects of EV charging, national legislation has played a key role in the shaping of the new material political dynamics. First, national policies have been important for establishing Norway as a national EV laboratory (Skjølsvold and Ryghaug 2020), as discussed in Chap. 2. Here, however, we are more interested in the way that the national legislation has pushed EV charging from being a purely individual responsibility towards becoming recognized as a more collective challenge, and through this arguably orchestrating potential ways of participating beyond the individual. In early phases of the EV introduction, most EV owners in shared garage spaces simply used available electricity wall sockets. This resulted in increased fire hazards, and with time, also a lack of power capacity in some garages. The consequence was that many apartment buildings banned EV charging in shared spaces, thus effectively making EV ownership difficult. This made the headlines of local newspapers around the country, describing new conflicts between EV owners and elected boards of such neighbourhoods, signalling the emergence of a new form of local political battleground around the practices of EV charging.

These localized battles between EV owners and elected boards were echoed in parliament, where Norwegian politicians now realized that, in an unexpected way, the actions of these community boards had become what they perceived as a barrier to further advancing electromobility. Striving to counteract such movements, in 2018, parliament passed a new law, which essentially requires boards to facilitate for EV charging. Though this act, elected boards of such neighbourhoods were given the formal responsibility of standardizing and harmonizing charging solutions locally. Hence, the national legislation transformed elected housing boards into key actors in the production of future local energy systems.

Promoting Smart Charging Through Highlighting Values such as Fairness and Equality

The contemporary material politics of smart EV charging has been shaped in part through national regulations, and in part through experiences generated through interaction between individual homeowners, commercial actors and consulting electricians who work within the field. These encounters have been and continue to be central in shaping the material politics of smart charging. In the early phases of EV implementation in Norway, EV owners would contact professional actors to install private chargers. Many electricians and professional actors in this field, however, quite early recognized that these assignments had broader and systemic implications, and that there was a distinct political quality to the work they were doing. These actors articulated that the power capacity of shared garage spaces could be understood as a form of common good or 'a common pool resource' (Ostrom 1990; Wolsink 2012), which needed governance and regulation due to escalating peak loads and capacity problems. This meant that they recognized the fact that immediate individual needs and desires for installing electricity chargers should be suspended somewhat to cater for the needs of future collectives of citizens.

Amongst professional actors, such as consulting engineers, and especially electricians, it became an outspoken ambition to raise the awareness amongst housing boards and inhabitants related to needs concerning the electricity grid. Through these articulation processes, these actors were key to enabling local material politics of smart charging through actualizing the grid as an object of concern for more actors. Hence, they mediated and enrolled elected boards and citizens on behalf of the grid, which illustrates how material publics and material politics emerges through the work of a range of human and non-human actors (Chilvers and Kearnes 2015). A key point here is that this work did not primarily entail attempts to educate boards about technicalities of electricity grids and smart charging. Rather, the issue tended to be framed in terms of values and principles of equality and fairness that often was metaphorically linked to the political systems that different constellations of garages, electricity grids and electric vehicles were seen to emulate.

Thus, electricians who were operating in what has been defined as the second stage of the Norwegian EV transition (Ryghaug and Skjølsvold 2019) where EVs became mainstream expressed this sort of care, both concerning the grid, and for the communities served by the grid. For them, implementing a planned and structured system was considered a key to realizing equity and fairness, which were framed as being the opposite strategy of "infrastructural anarchy". Consulting engineers and electricians who worked either for their own firms or the electricity industry saw themselves as promoters of such ideals and worked actively to enrol other actors such car dealers in the promotion of planned and smart charging technology. These actors were key in orchestrating the participation of housing boards, who were enrolled primarily by the translation of material constraints and peak load electricity problems into tangible community concerns such as equity and fairness, which could be addressed through implementing smart charging.

Citizens and Elected Housing Boards Co-producing Material Participation

The actual implementation and standardization of such technologies is the responsibility of housing boards. Such boards tend to be comprised of individuals with a wide range of social backgrounds and competences. Through the influx of EVs the electricity grid is actualized as an object of concern for such management boards, as well as for many of the inhabitants of such neighbourhoods. One example of how the political dynamics might unfold can be found in the largest neighbourhood of this type in Norway, consisting of 1113 units. At the time of our study, the neighbourhood had installed 55 EV charging points. The adoption of EVs increased rapidly, however, and there was a strong pressure from residents to increase the capacity and to install more chargers. About 60 tenants in the community were on a waiting list for acquiring an EV charging point, and the board was frequently petitioned to increase the power capacity. Hence, EVs actualized an interest in engaging more actively with grid issues. After a lengthy dialogue with a local electrical engineer as well as the electricity provider and grid operator, the board decided to install 765 new smart EV charging points with a centralized load management system intended to minimize peak loads.

The board was convinced that introducing smart EV charging with load control would both ensure fair access to electricity and charging. Hence, they decided to work to anchor this decision amongst residents through voting over it at a general meeting. This illustrates how the board's sensitization to the grid had catered for an appreciation of the grid as a common pool resource (Wolsink 2012), and further, raised the need for producing a legitimate form of governance of electricity grids through smart charging.

First, they produced a vision in which smart charging was part of a broader set of infrastructural upgrades that would improve the quality of the urban residential community. Hence, making the area attractive for EVs was seen as a way to raise the value of the residential area in general. Further, the board invited a series of actors from the electricity sector to the meeting. These actors were mobilized to make the case that smart charging would be a good solution for the residents, but also to discuss what was the most suitable solution and the needs for charging with the residents. The electricity sector actors used the occasion to speak on behalf of the grid, and to articulate how the 'state of the grid' depended on the ways that the residents live their lives and use electricity. A key aspect here was illustrating that the grid was too weak to handle the new EV charging demand. The actors present at the meeting gave presentations that also concerned trends within electricity provision, expected developments in the future, as well as potential consequences for the urban residential communities of the different choices they could make collectively (installing shared smart charging vs. not installing). There was also elaborate room for questions and answers, which also extended more broadly into discussions about the role of urban residential communities in sustainability transitions.

In effect, the meeting became what Martiskainen et al. (2018) have called a material and discursive space of diverse voices, which enabled a new form of engagement with social and material aspects of the grid. This space, the presence of these diverse voices and forms of competences, as well as the materiality of EVs and the grid itself became part of an ecology (Chilvers et al. 2018) through which awareness, new knowledge and new practices could be constructed. Thus, one of the consequences were the increased attention to key aspects of energy use, and hence the emergence of a more collective form of energy citizenship than discussed earlier in this chapter. Hence, the influx of EVs into this urban residential community generated a new form of material politics, where sensitization to aspects of the electricity grid on behalf of a range of actors was key. The elected community board became champions for a relatively complex system of smart charging. The discussions about smart charging spilled over into discussions about equity and fairness, for example with respect to equal access to electricity for all across both time and space and the idea of capacity as a common good, but also about broader visions for what role this urban residential community should be in future society and energy system. Here, focus was that technologies such as smart charging could provide new benefits to the community. Finally, this process illustrates that material participation here, transgressed the domestic sphere where it has most frequently been analysed (e.g. Marres 2016; Ryghaug et al. 2019), to take the form of a formally organized democratic process through which a decision of implementing the technology was reached collectively.

Conclusion: Orchestration, Participation and New Collective Material Politics

In this chapter we have sought to open participation as a phenomenon within and around pilot and demonstration projects. In sum, our observations add to an understudied aspect of energy and mobility transition discussions. There has been a tendency in the research literature either to focus on aspects of pilot projects and demonstration projects that point towards top-down implementation, or to focus on bottom-up processes as a key mode of participation. These literatures are often linked to quite different discourses: on the one hand, the focus on top-down processes links to ideas about a neoliberal or post-political order, where control over critical infrastructure is moved from public and political institutions to private and economic actors under the pretence of empowering citizens or governments to make more active and better decisions. On the other hand, the focus on bottom-up lends itself to ideas about new forms of democratization, inclusion and decision making.

In contrast, our discussion in this chapter illustrates how bottom-up and top-down seemingly co-exist, in the same cases and in the same sites, but where focusing on the material objects provides new ways of seeing the formation of issues and publics. First, we have highlighted that the work to materialize visions of future technoscientific realities through making pilot and demonstration projects not only produces technologies that citizens can accept or reject, but that they inherently produce or orchestrate different modes of participation. A central aspect of this is that pilot and demonstration projects do so not only for citizens but for a range of actors. Hence, working to advance certain technoscientific realities is a form of material participation, which in turn produces conditions for the participation of others. This illustrates the way innovation is a political activity: it shapes our understanding of potential futures, it shapes our opportunities of acting within or towards such futures and it re-configures the roles and competences of a wide ecology of actors. In future research as well as for improving practice, it is central to understand better how such conditions are orchestrated, and how one can work to orchestrate the participation in more diverse and inclusive ways.

In this chapter we have been especially interested in the ways that objects and technologies might enable new forms of material participation, or what we have called energy citizenship. To recap, energy citizenship entails that engagement with objects are productive of new forms of awareness, new forms of literacy or knowledge, and the formation of new practices and actions with respect to energy issues. These traits of energy citizenship point towards the merit of relatively open forms of orchestration, that seeks to identify broader repertoires of action than those that are anchored in for example accept/reject dichotomies, or choices of consuming/not consuming. Seeking to build on traits such as learning, working together, building competence and new ways of acting seems promising, but there is a need for a much more systematic assessment of which modes of participation exists and how technologies can be produced and shaped to cater for this potential diversity.

Further, this chapter has hinted at how one can envision the formation of more systemic forms of material participation, which are enacted beyond individual households or beyond objects and "gadgets". We see great promise in this approach. For instance, we have discussed how the relationship between electric vehicles and the energy system has opened for a wider actualization of the electricity grid as a political object: one that can be mobilized and enrolled into a range of issues that perhaps do not intuitively stand out as related to EV charging. We have observed how such infrastructures, when made part of mundane discussions, indeed enable both awareness, knowledge and new practices and the way these articulations may be translated into participation both with respect to the energy system, and with respect to wider related political discussions about how to transform local communities into more fair or just communities with respect to energy.

In sum, this Chapter points towards a wide ecology of actors shaping participation, and the links between developing pilot projects and developing broader societal projects. This suggests an increasingly important role for social science and STS in deepening our understanding of and stressing the central role of science and technology in the making of democratic life (Laurent 2017; Jasanoff 2005); the implications of innovation and technology development for the orderings of society. While we have pointed towards some promising routes for thinking anew about the relationship between technology and participation, current dominant logics of innovation and technology development tends to (re-)produce technologies that open for very limited forms of participation. Technology development and innovation processes are still most often rooted in ideas about participation through market mechanisms and consumption, which means that as a whole, the current deployment of pilot and demonstration projects cannot necessarily be described as a democratic development. Rather than seeking to tweak or optimize this system, we believe there is a need to systematically re-think current research and innovation systems and policies. Research and innovation systems and policies should not only be more transparent and nurture new types of technologies. They should more actively seek out to nurture and experiment with social forms and more open forms of participation. This will be discussed in greater detail in the next chapter.

References

Akrich, M. (1992). The de-scription of technical objects. In W. Bijker & J. Law (Eds.), *Shaping technology/Building society: Studies in socio-technological change*. MIT Press.

Anfinsen, M., Lagesen, V. A., & Ryghaug, M. (2019). Green and gendered? Cultural perspectives on the road towards electric vehicles in Norway. *Transportation Research Part D: Transport and Environment, 71*, 37–46.

Ballo, I. F. (2015). Imagining energy futures: Sociotechnical imaginaries of the future Smart Grid in Norway. *Energy Research & Social Science, 9*, 9–20.

Barry, A. (2001). *Political machines: Governing a technological society*. A&C Black.

Borup, M., Brown, N., Konrad, K., & Van Lente, H. (2006). The sociology of expectations in science and technology. *Technology Analysis & Strategic Management, 18*(3–4), 285–298.

Callon, M. (1999). The role of lay people in the production and dissemination of scientific knowledge. *Science, Technology and Society, 4*(1), 81–94.

Canzler, W., Engels, F., Rogge, J. C., Simon, D., & Wentland, A. (2017). From "living lab" to strategic action field: Bringing together energy, mobility, and information technology in Germany. *Energy Research & Social Science, 27*, 25–35.

Chilvers, J., & Kearnes, M. (Eds.). (2015). *Remaking participation: Science, environment and emergent publics*. Routledge.
Chilvers, J., & Longhurst, N. (2016). Participation in transition (s): Reconceiving public engagements in energy transitions as co-produced, emergent and diverse. *Journal of Environmental Policy & Planning, 18*(5), 585–607.
Chilvers, J., Pallett, H., & Hargreaves, T. (2018). Ecologies of participation in socio-technical change: The case of energy system transitions. *Energy Research & Social Science, 42*, 199–210.
Christensen, T. H., Friis, F., Bettin, S., Throndsen, W., Ornetzeder, M., Skjølsvold, T. M., & Ryghaug, M. (2020). The role of competences, engagement, and devices in configuring the impact of prices in energy demand response: Findings from three smart energy pilots with households. *Energy Policy, 137*, 111142.
Delina, L. L., & Sovacool, B. K. (2018). Of temporality and plurality: An epistemic and governance agenda for accelerating just transitions for energy access and sustainable development. *Current Opinion in Environmental Sustainability, 34*, 1–6.
Devine-Wright, P. (2012). Energy citizenship: Psychological aspects of evolution in sustainable energy technologies. In *Governing technology for sustainability* (pp. 74–97). Routledge.
Felt, U., & Fochler, M. (2010). Machineries for making publics: Inscribing and de-scribing publics in public engagement. *Minerva, 48*(3), 219–238.
Fjellså, I., Silvast, A., & Skjølsvold, T. M. (forthcoming). Fair flexibility? Capabilities and framings of user flexibility in the electricity grid2. Under review in Environmental Innovations and Societal Transitions.
Geels, F. W. (2002). Technological transitions as evolutionary reconfiguration processes: A multi-level perspective and a case-study. *Research Policy, 31*(8–9), 1257–1274.
Geels, F. W. (2005). The dynamics of transitions in socio-technical systems: A multi-level analysis of the transition pathway from horse-drawn carriages to automobiles (1860–1930). *Technology Analysis & Strategic Management, 17*(4), 445–476.
Geels, F. W. (2010b). Ontologies, socio-technical transitions (to sustainability), and the multi-level perspective. *Research Policy, 39*(4), 495–510.
Heidenreich, S. (2015). Sublime technology and object of fear: Offshore wind scientists assessing publics. *Environment and Planning A, 47*(5), 1047–1062.
Ingeborgrud, L., Heidenreich, S., Ryghaug, M., Skjølsvold, T. M., Foulds, C., Robison, R., ... Mourik, R. (2020). Expanding the scope and implications of energy research: A guide to key themes and concepts from the Social Sciences and Humanities. *Energy Research & Social Science, 63*, 101398.
Ingeborgrud, L., & Ryghaug, M. (2019). The role of practical, cognitive and symbolic factors in the successful implementation of battery electric vehicles in Norway. *Transportation Research Part A: Policy and Practice, 130*, 507–516.

Irwin, A., & Wynne, B. (Eds.). (2003). *Misunderstanding science?: The public reconstruction of science and technology*. Cambridge University Press.
Jasanoff, S. (2005). Technologies of humility: Citizen participation in governing science. In *Wozu Experten?* (pp. 370–389). VS Verlag für Sozialwissenschaften.
Karlstrøm, H. (2012). Empowering markets? The construction and maintenance of a deregulated market for electricity in Norway.
Karlstrøm, H., & Ryghaug, M. (2014). Public attitudes towards renewable energy technologies in Norway. The role of party preferences. *Energy Policy, 67*, 656–663.
Kemp, R., Schot, J., & Hoogma, R. (1998). Regime shifts to sustainability through processes of niche formation: The approach of strategic niche management. *Technology Analysis & Strategic Management, 10*(2), 175–198.
Latour, B. (2005). From realpolitik to dingpolitik. In *Making things public: Atmospheres of democracy* (p. 1444). MIT Press.
Laurent, B. (2017). *Democratic experiments: Problematizing nanotechnology and democracy in Europe and the United States*. The MIT Press.
Marres, N. (2007). The issues deserve more credit: Pragmatist contributions to the study of public involvement in controversy. *Social Studies of Science, 37*(5), 759–780.
Marres, N. (2012). On some uses and abuses of topology in the social analysis of technology (or the problem with smart meters). *Theory, Culture & Society, 29*(4–5), 288–310.
Marres, N. (2013). Why political ontology must be experimentalized: On eco-show homes as devices of participation. *Social Studies of Science, 43*(3), 417–443.
Marres, N. (2016). *Material participation: Technology, the environment and everyday publics*. Springer.
Marres, N., & Lezaun, J. (2011). Materials and devices of the public: An introduction. *Economy and society, 40*(4), 489–509.
Martiskainen, M., Heiskanen, E., & Speciale, G. (2018). Community energy initiatives to alleviate fuel poverty: The material politics of Energy Cafés. *Local Environment, 23*(1), 20–35.
Ortar, N., & Ryghaug, M. (2019). Should all cars be electric by 2025? The electric car debate in Europe. *Sustainability, 11*(7), 1868.
Ostrom, E. (1990). *Governing the commons: The evolution of institutions for collective action*. Cambridge University Press.
Parag, Y., & Sovacool, B. K. (2016). Electricity market design for the prosumer era. *Nature Energy, 1*(4), 1–6.
Ryghaug, M., & Toftaker, M. (2014). A transformative practice? Meaning, competence, and material aspects of driving electric cars in Norway. *Nature and Culture, 9*(2), 146–163.

Ryghaug, M., & Skjølsvold, T. M. (2019). Nurturing a regime shift toward electro-mobility in Norway. In *The Governance of Smart Transportation Systems* (pp. 147–165). Cham: Springer.

Ryghaug, M., Sørensen, K. H., & Næss, R. (2011). Making sense of global warming: Norwegians appropriating knowledge of anthropogenic climate change. *Public Understanding of Science, 20*(6), 778–795.

Ryghaug, M., Skjølsvold, T. M., & Heidenreich, S. (2018). Creating energy citizenship through material participation. *Social Studies of Science, 48*(2), 283–303.

Ryghaug, M., Ornetzeder, M., Skjølsvold, T. M., & Throndsen, W. (2019). The role of experiments and demonstration projects in efforts of upscaling: an analysis of two projects attempting to reconfigure production and consumption in energy and mobility. *Sustainability, 11*(20), 5771.

Shapin, S., & Schaffer, S. (1985). *Leviathan and the air-pump: Hobbes, Boyle, and the experimental life* (Vol. 109). Princeton University Press.

Silvast, A. (2017). Energy, economics, and performativity: Reviewing theoretical advances in social studies of markets and energy. *Energy Research & Social Science, 34*, 4–12.

Skjølsvold, T. M. (2014). Back to the futures: Retrospecting the prospects of smart grid technology. *Futures, 63*, 26–36.

Skjølsvold, T. M., & Lindkvist, C. (2015). Ambivalence, designing users and user imaginaries in the European smart grid: Insights from an interdisciplinary demonstration project. *Energy Research & Social Science, 9*, 43–50.

Skjølsvold, T. M., & Ryghaug, M. (2020). Temporal echoes and cross-geography policy effects: Multiple levels of transition governance and the electric vehicle breakthrough. *Environmental Innovation and Societal Transitions, 35*, 232–240.

Skjølsvold, T. M., Throndsen, W., Ryghaug, M., Fjellså, I. F., & Koksvik, G. H. (2018). Orchestrating households as collectives of participation in the distributed energy transition: New empirical and conceptual insights. *Energy Research & Social Science, 46*, 252–261.

Strengers, Y. (2014). Smart energy in everyday life: are you designing for resource man?. *Interactions, 21*(4), 24–31.

Stilgoe, J., Lock, S. J., & Wilsdon, J. (2014). Why should we promote public engagement with science?. *Public Understanding of Science, 23*(1), 4–15.

Throndsen, W., & Ryghaug, M. (2015). Material participation and the smart grid: Exploring different modes of articulation. *Energy Research & Social Science, 9*, 157–165.

Throndsen, W., Skjølsvold, T. M., Ryghaug, M., & Christensen, T. H. (2017). From consumer to prosumer. Enrolling users into a Norwegian PV pilot. ECEEE Summer Study Proceedings, 2017.

Wallsten, A., & Galis, V. (2019). The discreet charm of activeness: The vain construction of efficient smart grid users. *Journal of Cultural Economy, 12*(6), 571–589.

Wolsink, M. (2012). The research agenda on social acceptance of distributed generation in smart grids: Renewable as common pool resources. *Renewable and Sustainable Energy Reviews, 16*(1), 822–835.

Wolsink, M. (2018). Social acceptance revisited: gaps, questionable trends, and an auspicious perspective. *Energy Research & Social Science, 46,* 287–295.

Wynne, B. (1996). A reflexive view of the expert-lay knowledge divide. In S. Lash, B. Szerszynski, & B. Wynne (Eds.), *Risk, environment and modernity: Towards a new ecology* (pp. 40–44). Sage.

Open Access This chapter is licensed under the terms of the Creative Commons Attribution 4.0 International License (http://creativecommons.org/licenses/by/4.0/), which permits use, sharing, adaptation, distribution and reproduction in any medium or format, as long as you give appropriate credit to the original author(s) and the source, provide a link to the Creative Commons licence and indicate if changes were made.

The images or other third party material in this chapter are included in the chapter's Creative Commons licence, unless indicated otherwise in a credit line to the material. If material is not included in the chapter's Creative Commons licence and your intended use is not permitted by statutory regulation or exceeds the permitted use, you will need to obtain permission directly from the copyright holder.

CHAPTER 4

Catering for Socio-technical Transformations: Rethinking Technology Policy for Inclusive Transformation

Abstract This chapter zooms out from looking at concrete pilot projects to looking more broadly at the implications of discussions on pilot projects as sites of politics. We discuss how such a perspective might feed into the work of innovators, funding bodies and the making of broader technology policy agendas. The chapter highlights the great potential in pilot projects as a mode of innovation for energy transitions, but bring to the fore the way such innovation activities often take on traditional and technology-centred characteristics. We argue that there is a need to change not only the ways that projects are funded to ensure diverse scientific participation. It is equally important to challenge the underlying assumptions and questions asked in pilot activities, as well as the goals of such energy transition activities. This entails a distributed agenda, where actors across the ecology of innovation share responsibilities for moving towards more just, democratic and humane modes of experimenting for sustainability.

Keywords Innovation policy • Innovation practice • Just transitions

In the previous chapters of this book, we have flagged the importance of pilot and demonstration projects as key activities in current energy and sustainability transitions. In this chapter we will zoom out from the projects and activities as such, to discuss a set of implications that follows from the three earlier chapters. Our discussion focuses on three distinct sites where our earlier analysis in this book is of relevance. First, we will look at implications for the types of pilot and demonstration projects that we have studied in this book. Second, we will look at the institutional embedding of such projects, primarily by discussing the relationship between research and innovation funding and possibilities of doing things differently within current frameworks. Third, we will discuss more broadly some challenges for contemporary innovation, research and technology policy in working to produce pilot and demonstration projects that not only cater for technology development, but also seek to achieve what we broadly can call a more inclusive transition.

These three sites are closely interlinked, constituting a wide ecology of actors, organizations and technologies that shape and actively orchestrate the potential roles that pilot and demonstration projects take on in sustainability transitions. Our key argument in this concluding chapter is that while piloting and demonstrations hold great potential as a mode of innovation for increased sustainability, there is still a need for transforming such activities to make them more inclusive and more oriented towards broader collectives of actors. Observations highlighting that this should be the collective responsibility of a wide ecology of actors, also opens for diagnosing contemporary innovation policy and suggests the development of somewhat different policy measures than those that are usually prescribed for improving innovation for sustainability.

Transforming the Innovation Practices of Pilot Projects

Through this book, we have argued that pilot and demonstration projects are a central mode of innovation for realizing many climate and sustainability ambitions. Beyond this, we have suggested that such projects are political entities: they do not only discretely change technological systems; they are potentially transformative for societies more broadly. This is one of the reasons why a reflective approach to participation is important, not as a matter of securing social acceptance for new technological configurations, but rather as a matter of securing legitimacy and opening up the issues within and around such projects to forms of democratic governance.

In this book, our discussions have primarily circled around innovation within the field of smart energy technology and electromobility. While our discussions have illustrated diversity on behalf of pilot and demonstration projects in terms of the degree of engaging socio-technical complexity, such projects also tend to re-produce what Chilvers et al. (2018) describe as dominant participatory collectives. In Chap. 3 we discussed participation as an orchestrated phenomenon (Skjølsvold et al. 2018). Keeping with this metaphor, we might say that the actors who set up and establish pilot projects tends to orchestrate participation either as a form of *consumption*, that they seek participation though igniting *behaviour change* or that they open for participation in the form of *consultation*.

On the one hand, participation orchestrated in these dominant ways limits potential transition agency on behalf of various publics, keeping it within well-defined and previously established roles that do not necessarily challenge the non-sustainable traits of contemporary societies. On the other hand, the technologies also tend to be framed relatively modestly, in the sense that the impacts of technologies are discussed in very narrow terms. As an example, experimenting with demand side management or demand-response technologies tends to be understood as very discretely relating to energy demand and its potential flexibility, in such a way that social agency is reduced to consumption (Wallsten and Galis 2019; Throndsen and Ryghaug 2015). Such technologies, however, have much broader potential social implications. From a critical perspective they might feed into social processes of re-producing energy poverty, traditional gender roles and other forms of inequalities (Suboticki et al. 2019; Powells and Fell 2019), but there are also examples of such technologies becoming catalysts of what we have called energy citizenship (Ryghaug et al. 2018).

Hence, we are faced with a situation in which many pilot and demonstration projects within smart energy and electro mobility mobilizes a very narrow conception both of what technology can do, and of what its social implications are. This is a paradox, because smart energy technologies tend to be part of elaborate visions not only of small energy system demand-side changes, but also of broad energy system and society wide transformation (see e.g. Strengers 2013; Goulden et al. 2014; Skjølsvold 2014; Ballo 2015). Such visions are sometimes formulated by scholars, but just as often circulate through the rhetoric of policy and industry. A prominent example of this is the EU strategic energy technology (SET) plan and its integrated policy roadmap, which state that activating and engaging consumers is a chief energy policy challenge in the EU for the coming years. Why are such sentiments not well reflected in the innovation practice of most pilot projects?

From our perspective, a key part of the problem at hand is that pilot and demonstration projects within smart energy technologies and electro mobility often have very clear technological goals, or what Weiland et al. (2017) call 'target knowledge'. Examples can be the integration of new renewable energy technologies or the balancing of supply and demand in the electricity system. On the other hand, the social goals, or social target knowledge tends to be formulated weakly, for instance as an abstract and non-specific idea of becoming flexible energy users or more active consumers.

Thus, a key challenge for innovators who engage in the production of pilot and demonstration projects is to symmetrically develop both social and technical target knowledge associated with their projects. Over the last decades, the social sciences and humanities have produced a vast repertoire of potential forms of social target knowledge (see. e.g. Ingeborgrud et al. 2020; Sovacool 2014; Sovacool et al. 2020). Examples are manifold and rich within the literature on energy justice (e.g. Jenkins et al. 2018), where there has also been developed specific normative analytical categories for smart energy technologies, such as the goal of achieving flexibility justice (Powells and Fell 2019). To us, then, a key question remains: Why is target knowledge most often deeply anchored in very specific technological goals and seldom formulated in terms of visions about social factors and future societies? Why has it become common sense for research projects to highlight the need for technological progress within rather limited and narrow technological areas and to a lesser extent fund research that follow up on the societal transitions needed in order to meet future societal goals such as low carbon societies and sustainable living? Later in this chapter, we will relate this strongly to the role of funding agencies, but for now, we will continue to probe the logics within projects.

The Challenge of the Social: Socio-technical Asymmetries in Pilot and Demonstration Projects

Through decades of studies on design processes, scholars from STS have illustrated that technology developers mobilize insights, or imaginations of technology users as key resources in design and innovation processes (e.g. Woolgar 1990; Akrich 1992; Oudshoorn and Pinch 2003). Our discussions in Chaps. 2 and 3 illustrate that the same is true within smart energy technology and electromobility (see e.g. Skjølsvold and Ryghaug 2015; Ryghaug and Toftaker 2016). Those who develop pilot and demonstration projects like those we have discussed, often do so either though imagining themselves as the ideal users of their own technologies (Strengers 2013, 2014), or through imagining technology users as a potential hurdle which should ideally be bypassed through clever design or automation (Fjellså et al. forthcoming; Skjølsvold et al. 2019). We will return to questions concerning the role of funding mechanisms later.

At this point we will note that there is a gap between the rhetoric of European policy, and the practice of many pilot projects, where policy rhetoric points towards ambitious social target knowledge while this is not always followed up in practice. From our perspective this is both disappointing and a missed opportunity. In practice, such asymmetries can take on many forms. In some instances, demonstration projects might masquerade a pure focus on technology with rhetoric on user involvement and user centric design in order to please funders or other external stakeholders that have attempted to orchestrate the participation of technology users as co-designers. To provide one example, this was observed in a large European interdisciplinary research and innovation project (Skjølsvold and Lindkvist 2015) where the technological goals of the project were clear; while leaving other goals in the dark. On the one hand, the project sought to verify technically that households in Italy and Germany were in principle able to produce as much electricity as they needed from newly installed solar panels. However, actors within this project had identified the mismatch between production and demand from solar power as a challenge, and hence intended to implement smart energy management as a way of exchanging electricity between buildings in a neighbourhood, in order to ensure that supply and demand were matched.

The project flagged ambitions of such neighbourhoods becoming entities of sharing and collaborating, hence quite clearly articulating a set of socially oriented forms of target knowledge for the pilot project. Given our discussions on material participation and energy citizenship in Chap. 3 (see also Ryghaug et al. 2018), one might intuitively consider this project as one that, on the one hand, tested a set of new technologies, while, on the other hand, using that technology to raise a set of public concerns or issues, allowing for new forms of collaborative participation in transition activities on behalf of the involved households. This might comprise a quite radical orchestration of public participation through smart energy technologies. As a way of achieving these quite ambitions social targets, the actors behind this project formulated ambitions of co-designing the involved smart energy technologies together with the prospective technology users of the project.

If taken at face value, this large European project involving several demonstration sites, explicitly relied on users participating as designers to achieve success. In practice, things looked different. From the start, the involved researchers and innovators were very sceptical towards involving users in real design and development activities. The scepticism was often rooted in questions concerning whether users really could understand such complex technologies in ways that made it possible for them to participate in technology design exercises. Through a gradual process, the goals of co-design were dismissed in practice. The rhetoric of co-design remained, however, and a form of co-design workshops was conducted. No actual users were invited to these workshops, and instead, project engineers *acted* as users themselves. This was justified by the fact that in principle, they could very well become future users of the technologies they were developing.

This story serves to illustrate that while the social aspects of many energy pilot and demonstration projects today makes up a much larger share of the narratives of such projects, these narratives do not necessarily reflect actual project practice. On one level, this might reflect a status gap in the relationship between social and technical knowledge within such projects (Ingeborgrud et al. 2020). Just as important, however, is probably the relationship between individual projects such as this, and their funders. Gram-Hanssen and Darby (2018) point out that funders who ask for citizen participation and engagement tend to do so quite vaguely, and in ways that open for purely tokenistic acts of inclusion. A possible interpretation of this, building on the work of Colette Bos et al. (2014), is that European science policy and funding mechanisms tend to cyclically mobilize new terms within their documents, without necessarily imposing strong sanctions on actual science and innovation practice. In what follows we will dive deeper into the relationship between funding and innovation practice, and potential consequences for citizen participation and inclusion.

THE ORCHESTRATION OF RESEARCH AND INNOVATION THROUGH FUNDING

Over the last pages, we have critically discussed some traits that we find problematic in many contemporary pilot and demonstration activities. Keeping with a perspective where we see transitions as enacted by a wide array of actors, changing this situation is not the sole responsibility of individual researchers or consortia. Changing the dynamics of the way research and innovation is conducted and the logics of such activities are also, of course, a form of socio-technical and multi-level transition. As we discussed in Chap. 1, the dynamics of such transitions entails work amongst a wide array of different actors engaged and embedded in various contexts. It also takes time.

As we hinted at in Chap. 2, access to research and innovation funding is a key resource mobilized in the making of pilot and demonstration projects. Such funding is often a way for authorities of different kinds to enact innovation, climate, transport or energy policy. An example of this is the Norwegian government's efforts to implement "National centres for environmentally friendly energy" (called FMEs), which was a direct consequence of a cross parliamentary political agreement on how to tackle climate change. The outcomes were a series of long-term funded research centres, of which the vast majority have been technological in character. The explicit mandate was to engage in value creation and innovation within green energy technology in close cooperation with industry. This is a funding mechanism with some similarities to those of UK Energy Research Centre (UKERC) and other similar UK-based initiatives (Winskel 2018). A key activity across the Norwegian FME centres was the setting up of pilot and demonstration projects. As Schot and Steinmueller (2018) have highlighted it is very difficult to assure that such efforts go beyond classical technology-oriented pilots. Hence, this is one example of how governments work to orchestrate the work of researchers and technology developers, where the outcome quite predictably was the formation of a set of large consortia which primarily engages in classical technology development activities.

On a basic level, the Norwegian example of centres for environmentally friendly energy reflects scholarship highlighting that in pure shares of funding, the social sciences are vastly inferior to the technical and natural sciences within climate and energy research (e.g. Øverland and Sovacool 2020; Foulds and Robison 2018). Yet, our argument here is not that funding more social science would necessarily enable more radical and socially oriented target knowledge in future pilot and demonstration projects. Such projects are shaped not only by the funding but also by the underlying assumptions concerning the character of innovation, the character of human rationality and the roles that technologies are expected to play in future societies. This has been illustrated repeatedly in research on smart energy technologies (Skjølsvold 2014; Ballo 2015), and recently these dynamics have also been shown in analyses of specific funders (Foulds and Christensen 2017).

Through our discussions in Chaps. 2 and 3, we have seen that many pilot and demonstration projects rely on funding from the European Union, and specifically from Horizon 2020, which over the last years have been the key framework programme for supporting research and innovation within the EU. Within the domain of energy, European research and innovation funding has arguably marginalized research on demand side technology use and citizens at the expense of a focus on energy production technologies (Wilson et al. 2012; Foulds and Robison 2018). The increased focus on "accelerated energy innovation" that has become a prominent aspect of energy policy-making in response to more urgent need for change have probably also contributed to this development, as focus has been directed towards cost reductions and deployment support as well as a central role for the private sector and public-private partnerships in transitions (Winskel and Radcliffe 2014).

More recently, however, "active consumption" has become one of the pillars around which projects from this framework programme is funded, and indeed, from the sorts of pilot and demonstration projects we discussed in Chaps. 2 and 3 it is clear that there is often a focus on changing what happens at the demand-side of the energy system. An important question for us, then, is why this apparent turn in the logics of European research and innovation funding has not resulted in much clearer formulations of social target knowledge within pilot and demonstration activities across Europe?

Foulds and Christensen (2017) have provided some important clues, in their analysis of the assumptions that underpin funding of energy research in the Horizon 2020, and specifically the ways that this program conceptualizes the relationship between social and technological development (see also Foulds et al. 2019). According to these scholars, the energy working programme is firmly rooted within a techno-economic paradigm, which rests on a dual conceptualization of human agency. On the one hand, people are primarily understood to act in the capacity of being rational consumers of energy, which results in a focus on behaviour change, decision making and choice, in other words well within what Chilvers et al. (2018) refer to as 'dominant modes of participation'. The key strategies for making consumers more active include raising awareness and providing information (see also Throndsen and Ryghaug 2015). On the other hand, people are conceptualized as a potential non-technical barrier to the diffusion of new technologies (see e.g. Skjølsvold 2012; Eaton et al. 2017). Hence, citizens are, on the one hand, seen as enablers of the transition through active consumption, while on the other hand being considered a barrier through resisting and rejecting new technologies. Thus, Foulds and Christensen's (2017) policy analysis clearly reflects the tension earlier described and identified within one single project (Skjølsvold and Lindkvist 2015).

A similar story can be told about the transport domain. Here, a key mechanism for the EU's goal to diversify and strengthen energy options for sustainable transport is through the Strategic Energy Technology (SET) Plan that sets out to increase energy efficiency and speed up the decarbonization of the transport sector, mainly by boosting research and innovation. Two key actions are put forward for this: (1) Action 7—becoming competitive in the global battery sector to drive e-mobility forward; and, (2) Action 8—strengthening market take-up of renewable fuels needed for sustainable transport solutions (European Commission 2017; Ryghaug et al. 2019). Hence, the goals are purely technological, with the social world being reduced to "market uptake". Hence, this policy disregards the vast body of socio-technical knowledge on the diverse and comprehensive processes that are needed across transport and mobility systems to achieve the needed societal transition (Hopkins and Higham 2016; Suboticki et al. 2019).

All of this means that the way participation is orchestrated in individual projects is no coincidence. Contrarily, the orchestration of participation is tightly linked to broader repertoires of understandings of human rationality and technology diffusion that circulate through networks of policy makers, funders and research scholars. While there has clearly been an expansion in focus in terms of opening for energy research and innovation projects that includes a focus on people and human action, the types of questions asked, the goals formulated and the technologies developed still appear to be restricted and confined to them being embedded in a rather tight and limiting techno-economic paradigm. The result is, on the one hand, a quest to optimize current patterns of energy consumption and production, rather than questioning the logics residing behind such patterns and seeking radical alternatives. Further, funding mechanisms and call texts give little room to ask questions that either do not concern how to optimize behaviour or how to diffuse the production and deployment of new technologies.

Re-thinking Transformative Innovation in Inclusive, Material and Human Terms

Our discussion above points to a relationship between the making of pilot- and demonstration projects, and the ways that such projects are often funded and anchored in institutions that operate based on narrow definitions of human rationality and technological development. If we are to return to the language of the multi-level perspective (MLP) as discussed in Chap. 1 (e.g. Geels 2002), we might say that many of the developments emerge from within a regime of research and development, where these definitions and understandings of human behaviour and technological development are integral aspects of these regimes semi-coherent grammar or rule-set. Keeping with this perspective, this means that the changes emerging from such endeavours are likely to be incremental rather than transformative.

The sorts of criticisms that we have discussed above have deep roots in decades of work at the intersection of social sciences, technological sciences, natural sciences and innovation studies. During the 1990s, Constructive Technology Assessment (e.g. Schot 1992; Rip et al. 1995) was put forward, as a framework where social scientists would work as mediators, bridging separate worlds, and through this addressing societal concerns emerging around scientific practice and innovation activities. In the same period, scholars within STS (e.g. Gibbons 1994) advanced ideas of context sensitive, problem focussed and interdisciplinary research under the banner of 'mode 2 science' or socially distributed knowledge (Nowotny et al. 2013). Others flagged the merits of research incorporating values such as unpredictability, incomplete control and a plurality of legitimate perspectives as a response to emerging challenges such as climate change, often described as 'post-normal science' (Funtowicz and Ravetz 1995).

The acronyms ELSI and ELSA were also put to work during the 1990s, as a way of dealing with ethical, legal and social implications/aspects, mainly of the life sciences, genomics and associated innovation. While they became thematically significantly expanded over the years, these programs have also become criticized for maintaining quite narrow conceptualizations of relevant issues rooted in ethics and risks, while overlooking more fundamental and systemic issues such as global justice and environmental issues (Zwart et al. 2014).

Later, related ideas have been promoted under the banner of Responsible Research and Innovation (RRI) (Von Schomberg 2013; Stilgoe et al. 2013), where the goal has been to build anticipatory, inclusive, reflexive and responsive practices of science and innovation. Both ELSA and RRI have strived to increase collaborations between social scientists and researchers and innovators from the technical and natural sciences. While many scholars promoting such perspectives flag far reaching ambitions on improving the conditions for how to govern emerging technologies, both approaches have been criticized of being institutionalized mainly as hedging mechanisms, where discussions about social and ethical aspects of technology early in innovation processes is expected to secure acceptance and reduce resistance to the end products of innovation processes (e.g. Zwart et al. 2014).

With a basis in challenges such as climate change and sustainability, recent years have seen the emergence of much critical debate about contemporary innovation systems. Johan Schot and Laur Kanger (2018) have suggested that the dominant rules and Meta rules that have guided the last 250 years of modernization might be fundamentally at odds with achieving key sustainability goals. The role of science and technology within this process has been framed either as a vehicle of economic growth to enable mass production and consumption, or as input factor to national innovation systems that strive to produce domestic growth in a globally competitive landscape. Such systems have produced economic growth, but also the double challenge of environmental degradation and social inequality (Schot and Steinmueller 2018). As an alternative, they call for a new and transformative innovation policy, which starts from the assumption that *"innovation cannot be equated with social progress, even when corrective social policies are in place"* (ibid., p. 1562). This claim is made based on the observation that many high-tech developments fuel rising global inequalities and rests on assumptions of increased growth and natural resource degradation.

For these scholars, the sustainable development goals (SDGs) constitutes a set of challenges so fundamentally different from the challenges solved through technological innovations in the past, that one should challenge the notions that industry transformation or technology transition is enough. Instead, the notion of transformative innovation policy suggests that one should actively seek *socio-technical system transformation*. As an example, the authors use the systems of transportation. Here, the role of science, technology and innovation policy within ordinary innovation systems would be to improve battery capacity in order to electrify new domains of transportation. In this case, one could end up transforming industry structures to some extent and reducing direct CO_2 emissions, but one might not realize the SDGs. Instead, Schot and Steinmueller argue that innovation policy should be mobilized to:

> [...] Supporting the emergence of new mobility systems in which for example private car ownership is less important, other mobility modalities such as small taxi vans, public transportation, walking and bicycling are more used in combination with for example electric vehicles provided by types of companies dedicated to the provision of mobility services using ICT capabilities. In this new system, mobility planning and thus also reduction of mobility has become an objective of all actors, and even a symbol of modern behavior. (Schot and Steinmueller 2018, p. 1562)

Hence, they foreground social innovation, the need to explicitly formulate social goals rooted in notions of justice, to go alongside technological goals. This entails expanding on who is involved in innovation and highlighting that citizens, NGOs and marginalized groups should play active roles. A key element of their argument is that producing such an innovation system would not only entail formulating a new policy for innovation, but also opening for and cultivating spaces of contestation and politics. These politics would be founded on spaces of experimentation, societal learning, public debate, deliberation and negotiation, which would be impossible without broad societal participation.

Presently, there is a stream of scholarship which resonates well with these ideas, calling for increased attention to what such an inclusive and transformative form of innovation policy and politics might entail in practice. We find this encouraging, because these literatures target the sorts of tensions highlighted in Chap. 1, where we argued that studies rooted within the MLP have had a tendency to focus quite narrowly on technological innovation and diffusion, while studies emerging from STS have arguably had a broader focus on the role of technologies in society. An example, is Jenkins et al. (2018), call for stronger inclusion of ethics and justice, both within socio-technical analysis of transitions, and in policy making in order to make unfolding transitions more just. Further, Delina and Sovacool (2018) point to the merits of plurality and diversity both in terms of recognizing scientific insights across technical, natural and social sciences, and within organization and decision making in their call for more human and just transitions. Valkenburg (2020) notes that calls for plurality of voices in innovation governance tends to result in a focus on forming consensus but argues that a contestation-oriented order might be equally fruitful. Jenkins et al. (2020) have further formulated an agenda which, on the one hand, points to the scholarly merits of energy justice, and on the other hand, also targets the institutional landscape of academia, as well as the relationships between academia and the world around. Resonating well with our discussions above, they highlight the need to challenge dominant funding traditions, to find new ways of relating to non-academics, and to not only produce visions about just energy futures, but to work actively to translate these into practice as well as highlighting whose responsibilities it is to effectuate these practices.

Jasanoff (2018) formulates an agenda that resembles that promoted by Johan Schot and his colleagues working on transformative innovation policy. Like Schot and Kanger (2018), she traces the roots of contemporary sustainability challenges to early enlightenment thinking and an ever-growing confidence that science and innovation can resolve all problems. Current science and innovation geared towards producing low-carbon energy futures, she points out, are formulated in very narrow ways, and are typically formulated by the already well-off. In working to advance transitions in the years ahead, she suggests mobilizing four "technologies of humility", which are intended both to sensitize us to the relationship between problems in the world and policy, and to humanize unfolding transitions. These technologies are: (1) Framing, that is, the foundations and focus of our scientific endeavours and innovation activities. As an example, should we focus on improving the physical properties of energy systems, or should we rather emphasize improving the lives of those disrupted by global change? (2) Vulnerability, especially through a focus on how vulnerabilities are shaped by history, place, class and social connectedness (3) Distribution, which entails asking questions about how policies and innovation affect countries, regions and people differently, and how to bring the voices of those that tend to be marginalized into the expert-heavy negotiations about which energy futures to embark on or promote (4) Learning, which she argues is currently constrained by the frames imposed on transition activity, and hence needs to be opened to ambiguity in order to make room for more thorough reflection on societal wide experience, and the strength and weaknesses of different approaches.

The above discussion signals a stream of contemporary critique against current innovation systems and innovation policies and their ability to produce the sorts of outcomes that are needed considering the overwhelming climate and sustainability challenges. We sympathize with these criticisms. Nevertheless, we have throughout this book discussed several pilot and demonstration projects, where we see significant potential for transformative change. In part, this can be attributed to the fact that the reach of innovation policy is not all encompassing, and that sometimes current dominant innovation policy and the rationalities within them can be enrolled and mobilized for other causes than those visible at face value. Further, we have seen examples of actors such as large companies, consulting electricians, citizens and policy makers who all in different ways take part in pilot and demonstration projects and that in doing so mobilize a wide repertoire of social, political, technological and economic goals and strategies to advance transitions. We have also seen how material elements might enable new forms of participation and new forms of politics. Hence, as we now turn to the final and concluding paragraphs of this book, we want to remain critical, while highlighting the hopeful potential of doing things differently.

Conclusion: Democratic Innovation for Inclusive Transformation

Through this book, we have shed light on the role of pilot and demonstration projects in emerging energy and mobility transitions. On the one hand, we have discussed how such activities emerge, which resources are mobilized in their production and which roles they take on both in transition processes and in wider societal changes. This entailed a focus on processes and practices of scaling up, and crucially, on the politics of such projects. Given that such projects are political in character we continued to discuss what this entailed, how such politics are enacted through processes of orchestration, and how such projects might open for new materialized forms of participation and energy citizenship. Building from a critical discussion on the relationship between innovation policy and pilot activities, we now turn to formulate a set of insights that can be gleaned from our discussions.

First, from a co-productionist and ecological perspective rooted in STS, it becomes clear that the responsibilities of achieving systemic transformations are distributed in society and amongst actors that are spread out across levels, sectors and domains of society. This means that transformative change requires innovation policies that target a broad set of issues, actors and societal domains. Second, this observation calls for very ambitious goals in terms of inclusively working to engage societal actors in transition work. This means looking beyond the types of actors that are presently active in building consortiums, developing, testing and using technologies, to actively ask who are implicated by the proposed developments and to work actively to amplify the voices heard and included in such activities.

Third, inclusivity not only entails the enrolment of large amounts of actors in the pursuit of specific technological goals. Inclusivity also entails looking beyond the dominant forms of knowledge, to mobilize insights from different disciplines as well as from citizens and other implicated actors when framing and formulating questions. Further, it entails working actively to mobilize epistemic and practical diversity in processes of innovation. Finally, it entails systematically pushing to ensure diverse outcomes. Fourth, and building on the latter point, is the importance of producing innovation policy and pilot projects that formulates distinctly social goals to go alongside the technological ambitions, or indeed to conduct pilots where experimenting with and transforming the social is the end goal. Such goals can relate, for example, to the use of technologies, but they could also be much broader, for example formulated as a vision for how policy and innovation will contribute to a more equitable and just society.

Fifth, participation and engagement are key social aspects of many pilot and demonstration projects and should be given priority. Working actively to orchestrate modes of participation that goes beyond the consumer role or accepting ready-made technologies should be prioritized. Sixth, and relatedly, the material elements mobilized in pilot and demonstration projects (e.g. individual technologies, infrastructures), are political. Innovators and designers could therefore embrace and experiment with how to exploit this fact rather than attempt to disguise technologies as neutral market devices or objective commodities.

Seventh, and building on the above points, innovation policy should embrace social and normative goals about the state of future societies, and work to nurture, stimulate and shield the emergence of organizations, movements and networks that promotes sustainable social change in similar ways that they currently nurture new technologies.

Eighth, the fact that the responsibility for enacting transitions is distributed, that a wide array of actors, and traditions of knowledge production will be involved, also means that transitions will have to come to terms with a wide range of interests and rationalities. This means conflict and controversy will be the norm, rather than the exception. Transition scholars have tended to emphasize the need to build shared visions, alignment and consensus in processes of transition. This will likely not be feasible if the end goal is transformative change. Experimenting with modes of engagement and participation that unties, reveals and cultivates conflicts could be a viable option.

In sum, the eight insights highlighted above are intended to illustrate a way forward from what Colette Bos (2014) pointed to as steering through big words. Big words can become empty signifiers, and the above is an attempt to make signifiers like 'inclusion' and 'democracy' more tangible in the context of energy transitions in general, and pilot and demonstration projects, in particular. As we move forward as scholars and practitioners, this list is also humbling. It illustrates the massive challenges ahead of us, while also pointing to some relatively concrete steps we could take and build further on in future energy transitions endeavours. Many futures are at stake, and many futures can be produced. Pilot and demonstration projects play a vital role in the shaping of our futures and will continue to do so for a long time. Which roles they will play, is still an open question, but also something for us as social scientist, STS and/or transition scholars and researchers to shape, contest, challenge and rethink.

References

Akrich, M. (1992). The de-scription of technical objects. In W. Bijker & J. Law (Eds.), *Shaping technology/Building society: Studies in socio-technological change*. MIT Press.

Ballo, I. F. (2015). Imagining energy futures: Sociotechnical imaginaries of the future Smart Grid in Norway. *Energy Research & Social Science, 9*, 9–20.

Bos, C., Walhout, B., Peine, A., & van Lente, H. (2014). Steering with big words: Articulating ideographs in research programs. *Journal of Responsible Innovation, 1*(2), 151–170.

Chilvers, J., Pallett, H., & Hargreaves, T. (2018). Ecologies of participation in socio-technical change: The case of energy system transitions. *Energy Research & Social Science, 42,* 199–210.

Delina, L. L., & Sovacool, B. K. (2018). Of temporality and plurality: An epistemic and governance agenda for accelerating just transitions for energy access and sustainable development. *Current Opinion in Environmental Sustainability, 34,* 1–6.

Eaton, W. M., Burnham, M., Hinrichs, C. C., & Selfa, T. (2017). Bioenergy experts and their imagined "obligatory publics" in the United States: Implications for public engagement and participation. *Energy Research & Social Science, 25,* 65–75.

European Commission. (2017). *The Strategic Energy Technology (SET) Plan.* Accessed from: https://setis.ec.europa.eu/sites/default/files/setis%20reports/2017_set_plan_progress_report_0.pdf

Fjellså, I., Silvast, A., & Skjølsvold, T. M. (forthcoming). Fair flexibility? Capabilities and framings of user flexibility in the electricity grid2. Under review in Environmental Innovations and Societal Transitions.

Foulds, C., & Robison, R. (2018). *Advancing energy policy.* Cham: Palgrave Pivot.

Funtowicz, S. O., & Ravetz, J. R. (1995). Science for the post normal age. In *Perspectives on ecological integrity* (pp. 146–161). Dordrecht: Springer.

Geels, F. W. (2002). Technological transitions as evolutionary reconfiguration processes: A multi-level perspective and a case-study. *Research Policy, 31*(8–9), 1257–1274.

Gibbons, M. (Ed.). (1994). *The new production of knowledge: The dynamics of science and research in contemporary societies.* Sage.

Goulden, M., Bedwell, B., Rennick-Egglestone, S., Rodden, T., & Spence, A. (2014). Smart grids, smart users? The role of the user in demand side management. *Energy Research & Social Science, 2,* 21–29.

Gram-Hanssen, K., & Darby, S. J. (2018). "Home is where the smart is"? Evaluating smart home research and approaches against the concept of home. *Energy Research & Social Science, 37,* 94–101.

Hopkins, D., & Higham, J. E. (Eds.). (2016). *Low carbon mobility transitions.* Goodfellow Publishers Ltd.

Ingeborgrud, L., Heidenreich, S., Ryghaug, M., Skjølsvold, T. M., Foulds, C., Robison, R., ... Mourik, R. (2020). Expanding the scope and implications of energy research: A guide to key themes and concepts from the Social Sciences and Humanities. *Energy Research & Social Science, 63,* 101398.

Jasanoff, S. (2018). Just transitions: A humble approach to global energy futures. *Energy Research & Social Science, 35,* 11–14.

Jenkins, K., Sovacool, B. K., & McCauley, D. (2018). Humanizing sociotechnical transitions through energy justice: An ethical framework for global transformative change. *Energy Policy, 117*, 66–74.

Jenkins, K. E., Stephens, J. C., Reames, T. G., & Hernández, D. (2020). Towards impactful energy justice research: Transforming the power of academic engagement. *Energy Research & Social Science, 67*, 101510.

Nowotny, H., Scott, P. B., & Gibbons, M. T. (2013). *Re-thinking science: Knowledge and the public in an age of uncertainty*. John Wiley & Sons.

Oudshoorn, N. E., & Pinch, T. (2003). *How users matter: The co-construction of users and technologies*. MIT Press.

Øverland, I., & Sovacool, B. K. (2020). The misallocation of climate research funding. *Energy Research & Social Science, 62*, 101349.

Powells, G., & Fell, M. J. (2019). Flexibility capital and flexibility justice in smart energy systems. *Energy Research & Social Science, 54*, 56–59.

Rip, A., Misa, T. J., & Schot, J. (Eds.). (1995). *Managing technology in society*. London: Pinter Publishers.

Ryghaug, M., Ornetzeder, M., Skjølsvold, T. M., & Throndsen, W. (2019). The role of experiments and demonstration projects in efforts of upscaling: an analysis of two projects attempting to reconfigure production and consumption in energy and mobility. *Sustainability, 11*(20), 5771.

Ryghaug, M., Skjølsvold, T. M., & Heidenreich, S. (2018). Creating energy citizenship through material participation. *Social Studies of Science, 48*(2), 283–303.

Ryghaug, M., & Toftaker, M. (2016). Creating transitions to electric road transport in Norway: The role of user imaginaries. *Energy Research & Social Science, 17*, 119–126.

Schot, J., & Kanger, L. (2018). Deep transitions: Emergence, acceleration, stabilization and directionality. *Research Policy, 47*(6), 1045–1059.

Schot, J., & Steinmueller, W. E. (2018). Three frames for innovation policy: R&D, systems of innovation and transformative change. *Research Policy, 47*(9), 1554–1567.

Schot, J. W. (1992). Constructive technology assessment and technology dynamics: The case of clean technologies. *Science, Technology, & Human Values, 17*(1), 36–56.

Skjølsvold, T. M. (2012). Publics in the pipeline. In N. Möllers & K. Zachmann (Eds.), *Past and present energy societies*. Bielefeld: Transcript Verlag.

Skjølsvold, T. M. (2014). Back to the futures: Retrospecting the prospects of smart grid technology. *Futures, 63*, 26–36.

Skjølsvold, T. M., & Ryghaug, M. (2015). Embedding smart energy technology in built environments: A comparative study of four smart grid demonstration projects. *Indoor and Built Environment, 24*(7), 878–890.

Skjølsvold, T. M., Fjellså, I. F., & Ryghaug, M. (2019). Det fleksible mennesket 2.0. *Norsk sosiologisk tidsskrift, 3*(03), 191–208.

Skjølsvold, T. M., & Lindkvist, C. (2015). Ambivalence, designing users and user imaginaries in the European smart grid: Insights from an interdisciplinary demonstration project. *Energy Research & Social Science, 9*, 43–50.

Skjølsvold, T. M., Throndsen, W., Ryghaug, M., Fjellså, I. F., & Koksvik, G. H. (2018). Orchestrating households as collectives of participation in the distributed energy transition: New empirical and conceptual insights. *Energy Research & Social Science, 46*, 252–261.

Sovacool, B., Hess, D. J., Amir, S., Geels, F. W., Hirsh, R., Medina, L. R., … Yearley, S. (2020). Sociotechnical agendas: Reviewing future directions for energy and climate research. *Energy Research and Social Science, 70*, 1–35.

Sovacool, B. K. (2014). What are we doing here? Analyzing fifteen years of energy scholarship and proposing a social science research agenda. *Energy Research & Social Science, 1*, 1–29.

Stilgoe, J., Owen, R., & Macnaghten, P. (2013). Developing a framework for responsible innovation. *Research Policy, 42*(9), 1568–1580.

Strengers, Y. (2013). *Smart energy technologies in everyday life*: Smart Utopia?. Cham: Springer.

Strengers, Y. (2014). Smart energy in everyday life: are you designing for resource man?. *Interactions, 21*(4), 24–31.

Suboticki, I., Świątkiewicz-Mośny, M., Ryghaug, M., & Skjølsvold, T. M. (2019). *Inclusive engagement in Energy with special focus on low carbon transport solutions. Scoping workshop report.* Cambridge: Energy-SHIFTS.

Throndsen, W., & Ryghaug, M. (2015). Material participation and the smart grid: Exploring different modes of articulation. *Energy Research & Social Science, 9*, 157–165.

Valkenburg, G. (2020). Consensus or contestation: Reflections on governance of innovation in a context of heterogeneous knowledges. Science, *Technology and Society*, https://doi.org/10.1177/0971721820903005.

Von Schomberg, R. (2013). A vision of responsible research and innovation. In *Responsible innovation: Managing the responsible emergence of science and innovation in society* (pp. 51–74). Wiley.

Wallsten, A., & Galis, V. (2019). The discreet charm of activeness: The vain construction of efficient smart grid users. *Journal of Cultural Economy, 12*(6), 571–589.

Weiland, S., Bleicher, A., Polzin, C., Rauschmayer, F., & Rode, J. (2017). The nature of experiments for sustainability transformations: A search for common ground. *Journal of Cleaner Production, 169*, 30–38.

Wilson, C., Grubler, A., Gallagher, K. S., & Nemet, G. F. (2012). Marginalization of end-use technologies in energy innovation for climate protection. *Nature Climate Change, 2*(11), 780–788.

Winskel, M. (2018). The pursuit of interdisciplinary whole systems energy research: Insights from the UK Energy Research Centre. *Energy Research & Social Science, 37*, 74–84.

Winskel, M., & Radcliffe, J. (2014). The rise of accelerated energy innovation and its implications for sustainable innovation studies: A UK perspective. *Science & Technology Studies, 27*, 8–33.

Woolgar, S. (1990). Configuring the user: The case of usability trials. *The Sociological Review, 38*(1_suppl), 58–99.

Zwart, H., Landeweerd, L., & Van Rooij, A. (2014). Adapt or perish? Assessing the recent shift in the European research funding arena from 'ELSA' to 'RRI'. *Life Sciences, Society and Policy, 10*(1), 1–19.

Open Access This chapter is licensed under the terms of the Creative Commons Attribution 4.0 International License (http://creativecommons.org/licenses/by/4.0/), which permits use, sharing, adaptation, distribution and reproduction in any medium or format, as long as you give appropriate credit to the original author(s) and the source, provide a link to the Creative Commons licence and indicate if changes were made.

The images or other third party material in this chapter are included in the chapter's Creative Commons licence, unless indicated otherwise in a credit line to the material. If material is not included in the chapter's Creative Commons licence and your intended use is not permitted by statutory regulation or exceeds the permitted use, you will need to obtain permission directly from the copyright holder.

Correction to: Pilot Society and the Energy Transition

CORRECTION TO:

M. Ryghaug, T. M. Skjølsvold, *Pilot Society and the Energy Transition*, https://doi.org/10.1007/978-3-030-61184-2

The original version of this book was revised. The book was inadvertently published with an incorrect surname to Tomas Moe Skjølsvold of captioned title as "Moe Skjølsvold" whereas it should be "Skjølsvold". The author's name has been updated in the book. The correction to this book can be found at https://doi.org/10.1007/978-3-030-61184-2_5

The updated version of the book can be found at
https://doi.org/10.1007/978-3-030-61184-2

© The Author(s) 2021
M. Ryghaug, T. M. Skjølsvold, *Pilot Society and the Energy Transition*, https://doi.org/10.1007/978-3-030-61184-2_5

References

Åhman, M. (2006). Government policy and the development of electric vehicles in Japan. *Energy Policy 34*(4), 433–443.

Akrich, M. (1992). The de-scription of technical objects. In W. Bijker & J. Law (Eds.), *Shaping technology/Building society: Studies in socio-technological change*. MIT Press.

Åm, H. (2015). The sun also rises in Norway: Solar scientists as transition actors. *Environmental Innovation and Societal Transitions, 16*, 142–153.

Andersen, O. (2013). Towards the Use of Electric Cars. In *Unintended Consequences of Renewable Energy* (pp. 71–80). London: Springer.

Anfinsen, M., Lagesen, V. A., & Ryghaug, M. (2019). Green and gendered? Cultural perspectives on the road towards electric vehicles in Norway. *Transportation Research Part D: Transport and Environment, 71*, 37–46.

Asphjell, A., Asphjell, Ø., & Kvisle, H. (2013). *Elbil på Norsk*. Oslo: Transnova.

Ballo, I. F. (2015). Imagining energy futures: Sociotechnical imaginaries of the future Smart Grid in Norway. *Energy Research & Social Science, 9*, 9–20.

Barnett, J., Burningham, K., Walker, G., & Cass, N. (2012). Imagined publics and engagement around renewable energy technologies in the UK. *Public Understanding of Science, 21*(1), 36–50.

Barry, A. (2001). *Political machines: Governing a technological society*. A&C Black.

Berker, T., & Throndsen, W. (2017). Planning story lines in smart grid road maps (2010–2014): Three types of maps for coordinated time travel. *Journal of Environmental Policy & Planning, 19*(2), 214–228.

Bijker, W. E., Hughes, T. P., & Pinch, T. J. (Eds.). (1987). *The social construction of technological systems: New directions in the sociology and history of technology*. MIT Press.

Borup, M., Brown, N., Konrad, K., & Van Lente, H. (2006). The sociology of expectations in science and technology. *Technology Analysis & Strategic Management, 18*(3–4), 285–298.

Bos, C., Walhout, B., Peine, A., & van Lente, H. (2014). Steering with big words: Articulating ideographs in research programs. *Journal of Responsible Innovation, 1*(2), 151–170.

Broto, V. C., & Bulkeley, H. (2013). Maintaining climate change experiments: Urban political ecology and the everyday reconfiguration of urban infrastructure. *International Journal of Urban and Regional Research, 37*(6), 1934–1948.

Buchanan, R. (1992). Wicked problems in design thinking. *Design Issues, 8*(2), 5–21.

Bulkeley, H. A., Broto, V. C., & Edwards, G. A. (2014). *An urban politics of climate change: Experimentation and the governing of socio-technical transitions*. Routledge.

Buland, T. (1994). Framtiden er elektrisk. *IFIM-notat, 4*, 94.

Callon, M. (1984). Some elements of a sociology of translation: Domestication of the scallops and the fishermen of St Brieuc Bay. *The Sociological Review, 32*(1_suppl), 196–233.

Callon, M. (1999). The role of lay people in the production and dissemination of scientific knowledge. *Science, Technology and Society, 4*(1), 81–94.

Canzler, W., Engels, F., Rogge, J. C., Simon, D., & Wentland, A. (2017). From "living lab" to strategic action field: Bringing together energy, mobility, and information technology in Germany. *Energy Research & Social Science, 27*, 25–35.

Cetina, K. K. (1995). Laboratory studies: The cultural approach to the study of science. In *Handbook of science and technology studies* (pp. 140–167). Los Angeles: Sage Publishing.

Chilvers, J., & Kearnes, M. (Eds.). (2015). *Remaking participation: Science, environment and emergent publics*. Routledge.

Chilvers, J., & Longhurst, N. (2016). Participation in transition (s): Reconceiving public engagements in energy transitions as co-produced, emergent and diverse. *Journal of Environmental Policy & Planning, 18*(5), 585–607.

Chilvers, J., Pallett, H., & Hargreaves, T. (2018). Ecologies of participation in socio-technical change: The case of energy system transitions. *Energy Research & Social Science, 42*, 199–210.

Christensen, T. H., Ascarza, A., & Throndsen, W. (2013). Country-specific factors for the development of household smart grid solutions: Comparison of the electricity systems, energy policies and smart grid R&D and demonstration projects in Spain, Norway and Denmark. IHSMAG Project report. Retrieved

May 28, 2020, from https://vbn.aau.dk/ws/files/168269618/Christensen_et_al._Country_specific_factors_2013.pdf.
Christensen, T. H., Friis, F., Bettin, S., Throndsen, W., Ornetzeder, M., Skjølsvold, T. M., & Ryghaug, M. (2020). The role of competences, engagement, and devices in configuring the impact of prices in energy demand response: Findings from three smart energy pilots with households. *Energy Policy, 137,* 111142.
Coenen, L., Raven, R., & Verbong, G. (2010). Local niche experimentation in energy transitions: A theoretical and empirical exploration of proximity advantages and disadvantages. *Technology in Society, 32*(4), 295–302.
Collantes, G., & Sperling, D. (2008). The origin of California's zero emission vehicle mandate. *Transportation Research Part A: Policy and Practice, 42*(10), 1302–1313.
Collins, H. M. (1988). Public experiments and displays of virtuosity: The core-set revisited. *Social Studies of Science, 18*(4), 725–748.
Cotton, M., & Devine-Wright, P. (2012). Making electricity networks "visible": Industry actor representations of "publics" and public engagement in infrastructure planning. *Public Understanding of Science, 21*(1), 17–35.
Curtis, M., Torriti, J., & Smith, S. T. (2018). Demand side flexibility and responsiveness: Moving demand in time through technology. In *Demanding energy* (pp. 283–312). Cham: Palgrave Macmillan.
Delina, L. L., & Sovacool, B. K. (2018). Of temporality and plurality: An epistemic and governance agenda for accelerating just transitions for energy access and sustainable development. *Current Opinion in Environmental Sustainability, 34,* 1–6.
Devine-Wright, P. (2012). Energy citizenship: Psychological aspects of evolution in sustainable energy technologies. In *Governing technology for sustainability* (pp. 74–97). Routledge.
Eaton, W. M., Burnham, M., Hinrichs, C. C., & Selfa, T. (2017). Bioenergy experts and their imagined "obligatory publics" in the United States: Implications for public engagement and participation. *Energy Research & Social Science, 25,* 65–75.
Engels, F., & Münch, A. V. (2015). The micro smart grid as a materialised imaginary within the German energy transition. *Energy Research & Social Science, 9,* 35–42.
ETIP SNET. (2018). Vision 2050: Integrating smart networks for the energy transition: Serving Society and Protecting the Environment. Retrieved March 22, 2020, from https://www.etip-snet.eu/wp-content/uploads/2018/06/VISION2050-DIGITALupdated.pdf
European Commission. (2017). *The Strategic Energy Technology (SET) Plan.* Accessed from: https://setis.ec.europa.eu/sites/default/files/setis%20reports/2017_set_plan_progress_report_0.pdf

European Commission. (2020). *Horizon 2020*. Work Programme. Brussels. Accessed from: https://ec.europa.eu/research/participants/data/ref/h2020/wp/2018-2020/main/h2020-wp1820-intro_en.pdf

Evans, J., & Karvonen, A. (2011). Living laboratories for sustainability: Exploring the politics and epistemology of urban transition. In *Cities and low carbon transitions* (pp. 126–141). London: Routledge.

Evans, J., Karvonen, A., & Raven, R. (Eds.). (2016). *The experimental city*. Routledge.

Felt, U., & Fochler, M. (2010). Machineries for making publics: Inscribing and de-scribing publics in public engagement. *Minerva, 48*(3), 219–238.

Fjellså, I., Silvast, A., & Skjølsvold, T. M. (forthcoming). Fair flexibility? Capabilities and framings of user flexibility in the electricity grid2. Under review in Environmental Innovations and Societal Transitions.

Forlano, L. (2019). Cars and contemporary communications| Stabilizing/destabilizing the driverless city: Speculative futures and autonomous vehicles. *International Journal of Communication, 13*, 28.

Fosso, O. B., Molinas, M., Sand, K., & Coldevin, G. H. (2014). Moving towards the smart grid: The Norwegian case. In *2014 International Power Electronics Conference (IPEC-Hiroshima 2014-ECCE ASIA)* (pp. 1861–1867). IEEE.

Foulds, C., & Christensen, T. H. (2016). Funding pathways to a low-carbon transition. *Nature Energy, 1*(7), 1–4.

Foulds, C., & Robison, R. (2018). *Advancing energy policy*. Cham: Palgrave Pivot.

Frantzeskaki, N., Borgström, S., Gorissen, L., Egermann, M., & Ehnert, F. (2017). Nature-based solutions accelerating urban sustainability transitions in cities: Lessons from Dresden, Genk and Stockholm cities. In *Nature-based solutions to climate change adaptation in urban areas* (pp. 65–88). Cham: Springer.

Friis, F., & Christensen, T. H. (2016). The challenge of time shifting energy demand practices: Insights from Denmark. *Energy Research & Social Science, 19*, 124–133.

Frøysnes, A. S. (2014). Bare en jævla boks?: en analyse av visjonsarbeidet knyttet til avanserte måle- og styringssystemer (AMS) (Master's thesis, NTNU)

Funtowicz, S. O., & Ravetz, J. R. (1995). Science for the post normal age. In *Perspectives on ecological integrity* (pp. 146–161). Dordrecht: Springer.

Gangale, F., Vasiljevska, J., Covrig, C. F., Mengolini, A., & Fulli, G. (2017). *Smart grid projects outlook 2017*. The Netherlands, Petten: Joint Research Centre of the European Commission.

Geels, F. W. (2002). Technological transitions as evolutionary reconfiguration processes: A multi-level perspective and a case-study. *Research Policy, 31*(8–9), 1257–1274.

Geels, F. W. (2004). From sectoral systems of innovation to socio-technical systems: Insights about dynamics and change from sociology and institutional theory. *Research Policy, 33*(6–7), 897–920.

Geels, F. W. (2005). The dynamics of transitions in socio-technical systems: A multi-level analysis of the transition pathway from horse-drawn carriages to automobiles (1860–1930). *Technology Analysis & Strategic Management, 17*(4), 445–476.

Geels, F. W. (2006). Co-evolutionary and multi-level dynamics in transitions: The transformation of aviation systems and the shift from propeller to turbojet (1930–1970). *Technovation, 26*(9), 999–1016.

Geels, F. W. (2007). Feelings of discontent and the promise of middle range theory for STS: Examples from technology dynamics. *Science, Technology, & Human Values, 32*(6), 627–651.

Geels, F. W. (2010a). The role of cities in technological transitions: Analytical clarifications and historical examples. In H. Bulkeley, V. Castan Broto, M. Hodson, & S. Marvin (Eds.), *Cities and low carbon transitions* (pp. 29–44). Routledge.

Geels, F. W. (2010b). Ontologies, socio-technical transitions (to sustainability), and the multi-level perspective. *Research Policy, 39*(4), 495–510.

Geels, F. W. (2011). The multi-level perspective on sustainability transitions: Responses to seven criticisms. *Environmental Innovation and Societal Transitions, 1*(1), 24–40.

Geels, F. W., & Schot, J. (2007). Typology of sociotechnical transition pathways. *Research Policy, 36*(3), 399–417.

Geels, F. W., Sovacool, B. K., Schwanen, T., & Sorrell, S. (2017). Sociotechnical transitions for deep decarbonization. *Science, 357*(6357), 1242–1244.

Gibbons, M. (Ed.). (1994). *The new production of knowledge: The dynamics of science and research in contemporary societies.* Sage.

Goulden, M., Bedwell, B., Rennick-Egglestone, S., Rodden, T., & Spence, A. (2014). Smart grids, smart users? The role of the user in demand side management. *Energy Research & Social Science, 2*, 21–29.

Gram-Hanssen, K., & Darby, S. J. (2018). "Home is where the smart is"? Evaluating smart home research and approaches against the concept of home. *Energy Research & Social Science, 37*, 94–101.

Grydehøj, A., & Kelman, I. (2017). The eco-island trap: Climate change mitigation and conspicuous sustainability. *Area, 49*(1), 106–113.

Hargadon, A. B., & Douglas, Y. (2001). When innovations meet institutions: Edison and the design of the electric light. *Administrative Science Quarterly, 46*(3), 476–501.

Hargreaves, T., Longhurst, N., & Seyfang, G. (2013). Up, down, round and round: Connecting regimes and practices in innovation for sustainability. *Environment and Planning A, 45*(2), 402–420.

Haugland, B. T. (forthcoming) Self driving imaginaries, politics and innovation. Under review in Palgrave Communications.

Heidenreich, S. (2015). Sublime technology and object of fear: Offshore wind scientists assessing publics. *Environment and Planning A, 47*(5), 1047–1062.

Heiskanen, E., Hyvönen, K., Laakso, S., Laitila, P., Matschoss, K., & Mikkonen, I. (2017). Adoption and use of low-carbon technologies: Lessons from 100 Finnish pilot studies, field experiments and demonstrations. *Sustainability, 9*(5), 847.

Hess, D. J. (1997). *Science studies. An advanced introduction.* New York: New York University Press.

Hess, D. J., & Sovacool, B. K. (2020). Sociotechnical matters: Reviewing and integrating science and technology studies with energy social science. *Energy Research & Social Science, 65*, 101462.

Hopkins, D., & Higham, J. E. (Eds.). (2016). *Low carbon mobility transitions.* Goodfellow Publishers Ltd.

Hoogma, R. J., Kemp, R., Shot, J., & Truffer, B. (2002). *Experimenting for sustainable transport. The approach of strategic niche management.* Spon Press: London and New York.

Hughes, T. P. (1993). *Networks of power: Electrification in Western society, 1880–1930.* JHU Press.

Hughes, S., Chu, E. K., & Mason, S. G. (2018). *Climate change in cities.* Cham: Springer.

IEA. (2019). Renewables 2019, IEA, Paris. Retrieved from https://www.iea.org/reports/renewables-2019

Ingeborgrud, L., Heidenreich, S., Ryghaug, M., Skjølsvold, T. M., Foulds, C., Robison, R., ... Mourik, R. (2020). Expanding the scope and implications of energy research: A guide to key themes and concepts from the Social Sciences and Humanities. *Energy Research & Social Science, 63*, 101398.

Ingeborgrud, L., & Ryghaug, M. (2019). The role of practical, cognitive and symbolic factors in the successful implementation of battery electric vehicles in Norway. *Transportation Research Part A: Policy and Practice, 130*, 507–516.

Irwin, A., & Wynne, B. (Eds.). (2003). *Misunderstanding science?: The public reconstruction of science and technology.* Cambridge University Press.

Jasanoff, S. (Ed.). (2004). *States of knowledge: The co-production of science and the social order.* Routledge.

Jasanoff, S. (2005). Technologies of humility: Citizen participation in governing science. In *Wozu Experten?* (pp. 370–389). VS Verlag für Sozialwissenschaften.

Jasanoff, S. (2018). Just transitions: A humble approach to global energy futures. *Energy Research & Social Science, 35*, 11–14.

Jenkins, K., Sovacool, B. K., & McCauley, D. (2018). Humanizing sociotechnical transitions through energy justice: An ethical framework for global transformative change. *Energy Policy, 117*, 66–74.

Jenkins, K. E., Stephens, J. C., Reames, T. G., & Hernández, D. (2020). Towards impactful energy justice research: Transforming the power of academic engagement. *Energy Research & Social Science, 67*, 101510.

Karlstrøm, H. (2012). *Empowering markets? The construction and maintenance of a deregulated market for electricity in Norway.*

Karlstrøm, H., & Ryghaug, M. (2014). Public attitudes towards renewable energy technologies in Norway. The role of party preferences. *Energy Policy, 67*, 656–663.

Kårstein, A. (2008). *HyNor–den norskehydrogenveien?: En studie av en storteknopolitisk hybrid* (PhD Thesis, NTNU).

Katzeff, C., & Wangel, J. (2015). Social practices, households, and design in the smart grid. In *ICT innovations for sustainability* (pp. 351–365). Cham: Springer.

Kemp, R., Schot, J., & Hoogma, R. (1998). Regime shifts to sustainability through processes of niche formation: The approach of strategic niche management. *Technology Analysis & Strategic Management, 10*(2), 175–198.

Kemp, R. (2005). *Zero emission vehicle mandate in California: Misguided policy or example of enlightened leadership* (pp. 169–191). UK, Cheltenham: Edward Elgar.

Köhler, J., Geels, F. W., Kern, F., Markard, J., Onsongo, E., Wieczorek, A., ... Fünfschilling, L. (2019). An agenda for sustainability transitions research: State of the art and future directions. *Environmental Innovation and Societal Transitions, 31*, 1–32.

Kohler, R. E. (2002). Place and practice in field biology. *History of Science, 40*(2), 189–210.

Latour, B., & Woolgar, S. (1979). *Laboratory life: The construction of scientific facts.* Princeton University Press.

Latour, B. (1983). Give me a laboratory and I will raise the world. In K. Knorr-Cetina & M. Mulkay (Eds.), *Science observed: Perspectives on the social study of science* (pp. 141–170). London: Sage.

Latour, B. (1987). *Science in action: How to follow scientists and engineers through society.* Harvard University Press.

Latour, B. (1993). *The pasteurization of France.* Harvard University Press.

Latour, B. (2005). From realpolitik to dingpolitik. In *Making things public: Atmospheres of democracy* (p. 1444). MIT Press.

Latour, B., & Woolgar, S. (1979). *Laboratory life.* Beverly Hills, CA: Sage.

Laurent, B. (2017). *Democratic experiments: Problematizing nanotechnology and democracy in Europe and the United States.* The MIT Press.

Lie, M., & Sørensen, K. H. (Eds.). (1996). *Making technology our own?: Domesticating technology into everyday life.* Scandinavian University Press North America.

Markard, J., & Truffer, B. (2008). Technological innovation systems and the multi-level perspective: Towards an integrated framework. *Research Policy*, *37*(4), 596–615.

Marres, N. (2007). The issues deserve more credit: Pragmatist contributions to the study of public involvement in controversy. *Social Studies of Science*, *37*(5), 759–780.

Marres, N. (2012). On some uses and abuses of topology in the social analysis of technology (or the problem with smart meters). *Theory, Culture & Society*, *29*(4–5), 288–310.

Marres, N. (2013). Why political ontology must be experimentalized: On eco-show homes as devices of participation. *Social Studies of Science*, *43*(3), 417–443.

Marres, N. (2016). *Material participation: Technology, the environment and everyday publics*. Springer.

Marres, N. (2018). What if nothing happens? Street trials of intelligent cars as experiments in participation. In *TechnoScience in society, sociology of knowledge yearbook* (pp. 1–20). Nijmegen: Springer/Kluwer.

Marres, N. (2020). What if nothing happens? Street trials of intelligent cars as experiments in participation. In TechnoScienceSociety (pp. 111–130). Cham: Springer, Chicago.

Marres, N., Guggenheim, M., & Wilkie, A. (2018). *Inventing the social*. Manchester: Mattering Press.

Marres, N., & Lezaun, J. (2011). Materials and devices of the public: An introduction. *Economy and society*, *40*(4), 489–509.

Martiskainen, M., Heiskanen, E., & Speciale, G. (2018). Community energy initiatives to alleviate fuel poverty: The material politics of Energy Cafés. *Local Environment*, *23*(1), 20–35.

Mazzucato, M. (2015, January 2). Building the entrepreneurial state: A new framework for envisioning and evaluating a mission-oriented public sector. Levy Economics Institute of Bard College Working Paper No. 824. Retrieved from SSRN https://ssrn.com/abstract=2544707 or https://doi.org/10.2139/ssrn.2544707

Mazzucato, M. (2016). From market fixing to market-creating: A new framework for innovation policy. *Industry and Innovation*, *23*(2), 140–156.

Michael, M. (2000). Futures of the present. In N. Brown, B. Rappert, & A. Webster (Eds.), *Contested futures* (A sociology of prospective technoscience) (pp. 21–39). Aldershot: Ashgate.

Morozov, E., & Bria, F. (2018). *Rethinking the smart city*. New York: Rosa Luxemburg Stiftung.

Naber, R., Raven, R., Kouw, M., & Dassen, T. (2017). Scaling up sustainable energy innovations. *Energy Policy*, *110*, 342–354.

Nahuis, R., & Van Lente, H. (2008). Where are the politics? Perspectives on democracy and technology. *Science, Technology, & Human Values, 33*(5), 559–581.

Noel, L., de Rubens, G. Z., Sovacool, B. K., & Kester, J. (2019). Fear and loathing of electric vehicles: The reactionary rhetoric of range anxiety. *Energy Research & Social Science, 48,* 96–107.

Noel, L., Rubens, d., G. Z., Kester, J., & Sovacool, B. K. (2019). *Vehicle-to-grid: A Sociotechnical transition beyond electric mobility.* Springer.

Nowotny, H., Scott, P. B., & Gibbons, M. T. (2013). *Re-thinking science: Knowledge and the public in an age of uncertainty.* John Wiley & Sons.

Ortar, N., & Ryghaug, M. (2019). Should all cars be electric by 2025? The electric car debate in Europe. *Sustainability, 11*(7), 1868.

Ostrom, E. (1990). *Governing the commons: The evolution of institutions for collective action.* Cambridge University Press.

Oudshoorn, N. E., & Pinch, T. (2003). *How users matter: The co-construction of users and technologies.* MIT Press.

Øverland, I., & Sovacool, B. K. (2020). The misallocation of climate research funding. *Energy Research & Social Science, 62,* 101349.

Pallesen, T., & Jenle, R. P. (2018). Organizing consumers for a decarbonized electricity system: calculative agencies and user scripts in a Danish demonstration project. *Energy Research & Social Science, 38,* 102–109.

Park, S. (2011). Iceland's hydrogen energy policy development (1998–2007) from a sociotechnical experiment viewpoint. *International Journal of Hydrogen Energy, 36*(17), 10443–10454.

Papazu, I. (2016). Authoring participation. *Nordic Journal of Science and Technology Studies, 4*(1), 17–31.

Papazu, I. (2018). Storifying Samsø's renewable energy transition. *Science As Culture, 27*(2), 198–220.

Parag, Y., & Sovacool, B. K. (2016). Electricity market design for the prosumer era. *Nature Energy, 1*(4), 1–6.

Penna, C. C., & Geels, F. W. (2015). Climate change and the slow reorientation of the American car industry (1979–2012): An application and extension of the Dialectic Issue LifeCycle (DILC) model. *Research Policy, 44*(5), 1029–1048.

Pinch, T. J., & Bijker, W. E. (1984). The social construction of facts and artefacts: Or how the sociology of science and the sociology of technology might benefit each other. *Social Studies of Science, 14*(3), 399–441.

Powells, G., & Fell, M. J. (2019). Flexibility capital and flexibility justice in smart energy systems. *Energy Research & Social Science, 54,* 56–59.

Raven, R., Kern, F., Verhees, B., & Smith, A. (2016). Niche construction and empowerment through socio-political work. A meta-analysis of six low-carbon technology cases. *Environmental Innovation and Societal Transitions, 18,* 164–180.

Rip, A., Misa, T. J., & Schot, J. (Eds.). (1995). *Managing technology in society.* London: Pinter Publishers.

Roberts, C., Geels, F. W., Lockwood, M., Newell, P., Schmitz, H., Turnheim, B., & Jordan, A. (2018). The politics of accelerating low-carbon transitions: Towards a new research agenda. *Energy Research & Social Science, 44,* 304–311.

Rogelj, J., Shindell, D., Jiang, K., Fifita, S., Forster, P., Ginzburg, V. ... Vilariño, M. V. (2018). Mitigation pathways compatible with 1.5°C in the context of sustainable development. In: *Global Warming of 1.5°C. An IPCC Special Report on the impacts of global warming of 1.5°C above pre-industrial levels and related global greenhouse gas emission pathways, in the context of strengthening the global response to the threat of climate change, sustainable development, and efforts to eradicate poverty.* Retrieved May 26, 2020, from https://www.ipcc.ch/site/assets/uploads/sites/2/2019/02/SR15_Chapter2_Low_Res.pdf

Rosa, H. (2013). *Social acceleration: A new theory of modernity.* Columbia University Press.

Rosenow, J., & Kern, F. (2017). EU energy innovation policy: The curious case of energy efficiency. In *Research handbook on EU energy law and policy.* Edward Elgar Publishing.

Rosol, M., Béal, V., & Mössner, S. (2017). Greenest cities? The (post-) politics of new urban environmental regimes. *Environment and Planning A: Economy and Space, 49*(8), 1710–1718.

Rotmans, J., Kemp, R., & Van Asselt, M. (2001). More evolution than revolution: Transition management in public policy. *Foresight-The Journal of Future Studies, Strategic Thinking and Policy, 3*(1), 15–31.

Ryghaug, M., Ornetzeder, M., Skjølsvold, T. M., & Throndsen, W. (2019). The role of experiments and demonstration projects in efforts of upscaling: an analysis of two projects attempting to reconfigure production and consumption in energy and mobility. *Sustainability, 11*(20), 5771.

Ryghaug, M., & Skjølsvold, T. M. (2019). Nurturing a regime shift toward electro-mobility in Norway. In *The Governance of Smart Transportation Systems* (pp. 147–165). Cham: Springer.

Ryghaug, M., Skjølsvold, T. M., & Heidenreich, S. (2018). Creating energy citizenship through material participation. *Social Studies of Science, 48*(2), 283–303.

Ryghaug, M., & Sørensen, K. H. (2009). How energy efficiency fails in the building industry. *Energy Policy, 37*(3), 984–991.

Ryghaug, M., Sørensen, K. H., & Næss, R. (2011). Making sense of global warming: Norwegians appropriating knowledge of anthropogenic climate change. *Public Understanding of Science, 20*(6), 778–795.

Ryghaug, M., & Toftaker, M. (2014). A transformative practice? Meaning, competence, and material aspects of driving electric cars in Norway. *Nature and Culture, 9*(2), 146–163.

Ryghaug, M., & Toftaker, M. (2016). Creating transitions to electric road transport in Norway: The role of user imaginaries. *Energy Research & Social Science, 17*, 119–126.

Sadowski, J., & Levenda, A. M. (2020). The anti-politics of smart energy regimes. *Political Geography, 81*, 102202.

Schot, J., & Kanger, L. (2018). Deep transitions: Emergence, acceleration, stabilization and directionality. *Research Policy, 47*(6), 1045–1059.

Schot, J., Kanger, L., & Verbong, G. (2016). The roles of users in shaping transitions to new energy systems. *Nature Energy, 1*(5), 1–7.

Schot, J., & Steinmueller, W. E. (2018). Three frames for innovation policy: R&D, systems of innovation and transformative change. *Research Policy, 47*(9), 1554–1567.

Schot, J. W. (1992). Constructive technology assessment and technology dynamics: The case of clean technologies. *Science, Technology, & Human Values, 17*(1), 36–56.

Shapin, S., & Schaffer, S. (1985). *Leviathan and the air-pump: Hobbes, Boyle, and the experimental life* (Vol. 109). Princeton University Press.

Silvast, A. (2017). Energy, economics, and performativity: Reviewing theoretical advances in social studies of markets and energy. *Energy Research & Social Science, 34*, 4–12.

Skjølsvold, T. M. (2014). Back to the futures: Retrospecting the prospects of smart grid technology. *Futures, 63*, 26–36.

Silvast, A., Williams, R., Hyysalo, S., Rommetveit, K., & Raab, C. (2018). Who 'uses' smart grids? The evolving nature of user representations in layered infrastructures. *Sustainability, 10*(10), 3738.

Skjølsvold, T. M. (2012). Publics in the pipeline. In N. Möllers & K. Zachmann (Eds.), *Past and present energy societies*. Bielefeld: Transcript Verlag.

Skjølsvold, T. M., Fjellså, I. F., & Ryghaug, M. (2019). Det fleksible mennesket 2.0. *Norsk sosiologisk tidsskrift, 3*(03), 191–208.

Skjølsvold, T. M., & Lindkvist, C. (2015). Ambivalence, designing users and user imaginaries in the European smart grid: Insights from an interdisciplinary demonstration project. *Energy Research & Social Science, 9*, 43–50.

Skjølsvold, T. M., & Ryghaug, M. (2015). Embedding smart energy technology in built environments: A comparative study of four smart grid demonstration projects. *Indoor and Built Environment, 24*(7), 878–890.

Skjølsvold, T. M., & Ryghaug, M. (2020). Temporal echoes and cross-geography policy effects: Multiple levels of transition governance and the electric vehicle breakthrough. *Environmental Innovation and Societal Transitions, 35*, 232–240.

Skjølsvold, T. M., Ryghaug, M., & Berker, T. (2015). A traveler's guide to smart grids and the social sciences. *Energy Research & Social Science, 9*, 1–8.

Skjølsvold, T. M., Ryghaug, M., & Throndsen, W. (2020). European island imaginaries: Examining the actors, innovations, and renewable energy transitions of 8 islands. *Energy Research & Social Science, 65*, 101491.

Skjølsvold, T. M., Throndsen, W., Ryghaug, M., Fjellså, I. F., & Koksvik, G. H. (2018). Orchestrating households as collectives of participation in the distributed energy transition: New empirical and conceptual insights. *Energy Research & Social Science, 46*, 252–261.

Smith, A., & Raven, R. (2012). What is protective space? Reconsidering niches in transitions to sustainability. *Research Policy, 41*(6), 1025–1036.

Solbu, G. (2018). The physiology of imagined publics. *Science & Technology Studies, 31*, 39–54.

Sørensen, K. H. (2004). Cultural politics of technology: combining critical and constructive interventions?. *Science, technology, & human values, 29*(2), 184–190.

Sørensen, K. H. (2013). Beyond innovation. Towards an extended framework for analysing technology policy. *Nordic Journal of Science and Technology Studies, 1*(1), 12–23.

Sørensen, K. H., Lagesen, V. A., & Hojem, T. S. M. (2018). Articulations of mundane transition work among consulting engineers. *Environmental Innovation and Societal Transitions, 28*, 70–78.

Sovacool, B., Hess, D. J., Amir, S., Geels, F. W., Hirsh, R., Medina, L. R., ... Yearley, S. (2020). Sociotechnical agendas: Reviewing future directions for energy and climate research. *Energy Research and Social Science, 70*, 1–35.

Sovacool, B. K. (2014). What are we doing here? Analyzing fifteen years of energy scholarship and proposing a social science research agenda. *Energy Research & Social Science, 1*, 1–29.

Stilgoe, J., Owen, R., & Macnaghten, P. (2013). Developing a framework for responsible innovation. *Research Policy, 42*(9), 1568–1580.

Stilgoe, J., Lock, S. J., & Wilsdon, J. (2014). Why should we promote public engagement with science?. *Public Understanding of Science, 23*(1), 4–15.

Strengers, Y. (2013). *Smart energy technologies in everyday life: Smart Utopia?*. Cham: Springer.

Strengers, Y. (2014). Smart energy in everyday life: are you designing for resource man?. *Interactions, 21*(4), 24–31.

Suboticki, I., Świątkiewicz-Mośny, M., Ryghaug, M., & Skjølsvold, T. M. (2019). *Inclusive engagement in Energy with special focus on low carbon transport solutions. Scoping workshop report*. Cambridge: Energy-SHIFTS.

Throndsen, W. (2017). What do experts talk about when they talk about users? Expectations and imagined users in the smart grid. *Energy Efficiency, 10*(2), 283–297.

Throndsen, W., & Ryghaug, M. (2015). Material participation and the smart grid: Exploring different modes of articulation. *Energy Research & Social Science, 9*, 157–165.

Throndsen, W., Skjølsvold, T. M., Ryghaug, M., & Christensen, T. H. (2017). From consumer to prosumer. Enrolling users into a Norwegian PV pilot. ECEEE Summer Study Proceedings, 2017.

Tøndel, G., & Seibt, D. (2019). Governing the elderly body: Technocare policy and industrial promises of freedom. In *Digitalization in industry* (pp. 233–259). Cham: Palgrave Macmillan.

Torriti, J. (2015). *Peak energy demand and demand side response*. Chicago: Routledge.

Torriti, J. (2020). *Appraising the economics of smart meters: Costs and benefits*. Routledge.

Turnheim, B., Wesseling, J., Truffer, B., Rohracher, H., Carvalho, L., & Binder, C. (2018). Challenges ahead: Understanding, assessing, anticipating and governing foreseeable societal tensions to support accelerated low-carbon transitions in Europe. In *Advancing energy policy* (pp. 145–161). Cham: Palgrave Pivot.

Valkenburg, G. (2020). Consensus or contestation: Reflections on governance of innovation in a context of heterogeneous knowledges. Science, *Technology and Society*, https://doi.org/10.1177/0971721820903005.

Vasileiadou, E., & Safarzyńska, K. (2010). Transitions: Taking complexity seriously. *Futures, 42*(10), 1176–1186.

Vesnic-Alujevic, L., Breitegger, M., & Pereira, A. G. (2016). What smart grids tell about innovation narratives in the European Union: Hopes, imaginaries and policy. *Energy Research & Social Science, 12*, 16–26.

Von Schomberg, R. (2013). A vision of responsible research and innovation. In *Responsible innovation: Managing the responsible emergence of science and innovation in society* (pp. 51–74). Wiley.

Von Wirth, T., Fuenfschilling, L., Frantzeskaki, N., & Coenen, L. (2019). Impacts of urban living labs on sustainability transitions: Mechanisms and strategies for systemic change through experimentation. *European Planning Studies, 27*(2), 229–257.

Wallsten, A., & Galis, V. (2019). The discreet charm of activeness: The vain construction of efficient smart grid users. *Journal of Cultural Economy, 12*(6), 571–589.

Weiland, S., Bleicher, A., Polzin, C., Rauschmayer, F., & Rode, J. (2017). The nature of experiments for sustainability transformations: A search for common ground. *Journal of Cleaner Production, 169*, 30–38.

Wilkie, A., & Michael, M. (2018). Designing and doing: Enacting energy-and-community. In N. Marres, M. Guggenheim, & A. Wilkie (Eds.), *Inventing the social* (pp. 125–148). Manchester: Mattering Press.

Williams, R., & Edge, D. (1996). The social shaping of technology. *Research Policy, 25*(6), 865–899.

Wilson, C., Grubler, A., Gallagher, K. S., & Nemet, G. F. (2012). Marginalization of end-use technologies in energy innovation for climate protection. *Nature Climate Change, 2*(11), 780–788.

Winner, L. (1980). Do artifacts have politics? *Daedalus, 109*, 121–136.

Winskel, M. (2018). The pursuit of interdisciplinary whole systems energy research: Insights from the UK Energy Research Centre. *Energy Research & Social Science, 37*, 74–84.

Winskel, M., & Radcliffe, J. (2014). The rise of accelerated energy innovation and its implications for sustainable innovation studies: A UK perspective. *Science & Technology Studies, 27*, 8–33.

Wolsink, M. (2012). The research agenda on social acceptance of distributed generation in smart grids: Renewable as common pool resources. *Renewable and Sustainable Energy Reviews, 16*(1), 822–835.

Wolsink, M. (2018). Social acceptance revisited: gaps, questionable trends, and an auspicious perspective. *Energy Research & Social Science, 46*, 287–295.

Woolgar, S. (1990). Configuring the user: The case of usability trials. *The Sociological Review, 38*(1_suppl), 58–99.

Wynne, B. (1989). Sheepfarming after Chernobyl: A case study in communicating scientific information. *Environment: Science and Policy for Sustainable Development, 31*(2), 10–39.

Wynne, B. (1992). Misunderstood misunderstanding: Social identities and public uptake of science. *Public Understanding of Science, 1*(3), 281–304.

Wynne, B. (1996). A reflexive view of the expert-lay knowledge divide. In S. Lash, B. Szerszynski, & B. Wynne (Eds.), *Risk, environment and modernity: Towards a new ecology* (pp. 40–44). Sage.

Zwart, H., Landeweerd, L., & Van Rooij, A. (2014). Adapt or perish? Assessing the recent shift in the European research funding arena from 'ELSA' to 'RRI'. *Life Sciences, Society and Policy, 10*(1), 1–19.

Index

A
Acceleration, 10, 15, 45–47, 55
Accept, acceptance, 7, 13, 47, 64, 72, 75, 86, 87, 94, 103
Automation, 16, 32, 34, 80, 97
Awareness, 7, 76, 78, 83, 85, 87, 101

B
Battery, batteries, 16, 28, 40, 49, 77, 101, 104
Big words, 108
Business models, 5, 16, 17, 31

C
California, 39–41
Charging, charge, charger, 42, 77, 79–85, 87
Citizen, citizenship, 7, 34, 55, 65, 66, 69–72, 74–81, 83–87, 95, 97, 98, 100, 101, 104, 106, 107
Climate change, climate challenge, 7, 8, 10, 35, 56, 76, 78, 79, 99, 102, 103

Collectives, 10, 13, 26, 44, 47, 55, 67–74, 76–83, 85–88, 94, 95
Common pool resource, 16, 83, 85
Community, communities, 5, 30, 37, 38, 72, 73, 75, 78, 81–87
Consensus, 65, 105, 108
Constructive technology assessment, 102
Consumption, 14, 28, 33, 34, 44, 64, 68–71, 74, 78, 79, 88, 95, 100–103
Contestation, 11, 53, 73–74, 104
Cooling, 2, 15, 28
Co-produce, co-production, co-productionist, 13, 23–56, 65, 73, 76–79, 107

D
Demand, 2, 15, 16, 27, 28, 41–43, 47, 51, 66, 80, 85, 95–97, 100
Demand-side management, 70, 95
Democracy, democratic, democratization, 3, 6, 7, 11, 12, 17, 25, 34, 55, 56, 63–88, 94, 106–108

Demonstration, demonstrate, 1–17, 23–28, 30, 31, 33, 35–38, 43–48, 53–56, 63, 64, 66–72, 79, 80, 86, 88, 93–100, 102, 106–108
Design, 5, 32, 34, 66, 69, 71–74, 76, 96–98
Digitalisation, digital, 2, 27, 28

E
Ecology, ecologies, 7, 67, 73–74, 85, 87, 94
Economy, 39, 49, 50, 52
Electric vehicles (EVs), 7, 16, 28, 39–42, 72, 73, 76–85, 87, 104
Electrolysis, 49, 50
Electromobility, 4, 25, 39–42, 82, 94, 97
ELSA, 103
ELSI, 103
Energy citizenship, 7, 66, 71, 75–81, 85, 87, 95, 97, 107
Engagement, 11–13, 29, 44, 51, 65, 66, 68, 70, 72, 75–77, 79, 81, 85, 87, 98, 107, 108
Enrol, enrolment, 46, 47, 50, 67, 68, 70, 72, 84, 107
Epistemic culture, 30
European Commission (EU), 27, 29, 41, 95, 100
Europe, European, 5, 6, 15, 16, 24, 27, 30, 36–38, 45, 50, 66, 97, 98, 100
Experts, 25

F
Feedback, 16, 78, 80
Financial, 1, 27, 41, 42
Flexibility, 15, 16, 33, 69, 80, 95, 96
Fund, funding, 24, 26, 29, 30, 32, 36, 38, 40, 41, 44, 66, 70, 94, 96–102, 105

Future, 3, 4, 6, 8, 14, 16, 24–26, 28–30, 32, 35, 38, 41, 43, 44, 48–51, 53–55, 63, 66, 69–72, 78, 79, 82, 83, 85–87, 96, 98, 100, 105, 106, 108

G
Gas, 2, 14, 33
Gender, 95
Geography, geographical, 35, 36
Germany, 39, 97
Governance, 35, 48, 54, 55, 83, 85, 94, 105
Grid, 2, 6, 15, 16, 24, 25, 27–37, 42, 68, 80, 81, 83–85, 87

H
Health, healthcare, 33–35, 44, 68
Heating, 2, 15, 28, 32
Horizon 2020, 4, 24, 29, 36, 37, 100
Human, humane, 6, 7, 46, 67, 83, 100–106
Humility, 105
Hydrogen, 39, 49–52, 55
Hydropower, 40–42

I
ICT, 16, 34, 37, 44, 50, 54, 68, 69, 77, 104
Imagined, imaginary, imagined publics, 32, 37, 38, 49, 54, 66, 72
Inclusive, 13, 87, 93–108
Incremental, 9, 44, 102
Industry, 2, 12, 15, 16, 28–31, 33, 35, 38–41, 50, 69, 71, 73, 77, 84, 95, 99, 104
Infrastructure, 2, 10, 16, 29, 35, 42, 77, 81, 86, 87, 107

Innovation, 3–8, 10–14, 17, 24–30, 32, 34–38, 40, 43–46, 50, 51, 53–56, 64, 69–74, 76, 77, 79, 80, 87, 88, 94–108
Innovation system, 11, 88, 103, 104, 106
Institution, institutional, 2, 3, 6, 9, 10, 29, 33, 34, 47, 50, 52, 53, 68, 74, 77, 79, 80, 86, 94, 102, 105
Interessement, 46, 47, 49
Island, Islands, 36–38, 45, 69
Italy, 97

J
Japan, 40
Just, 3, 26, 41, 46, 50, 54, 64, 69, 80, 87, 95, 98, 105, 107
Justice, 11, 14, 35, 78, 96, 103–105

K
Knowledge, 7, 9, 47, 48, 64–66, 76, 78, 85, 87, 98, 101, 102, 107, 108

L
Laboratory, 4, 26–33, 35, 36, 38–42, 45, 46, 49–51, 68, 72, 82
Landscape, 9, 10, 72, 103, 105
Lay people, 5
Literacy, 7, 76, 78, 87
Living lab, 32
Load-shifting, 33

M
Material participation, 66, 71, 74–76, 78–80, 84–87, 97
Mobilization, 7, 25, 32, 33, 44, 46, 47, 64

Multi-level perspective (MLP), 8–14, 25, 30, 39, 53, 73, 102, 104
Municipalities, 35, 42, 52

N
Neighbourhoods, 31–33, 35, 45, 72, 81, 82, 84, 97
Network, 25, 35, 40, 42, 46, 47, 67, 71, 74, 81, 101, 108
Niche, 9, 10, 12, 39, 41, 64, 73, 81
Normative, norms, 8, 9, 64, 66, 73, 96, 108
Norway, 28–31, 33, 39–42, 48–50, 52, 72, 73, 81–84
Norwegian Research Council, 29

O
Objects, 6, 14, 67, 71, 74–76, 78, 79, 83, 84, 86, 87
Oil, 14, 33, 40
Orchestration, orchestrate, 66–74, 76, 86–88, 97, 99–102, 106

P
Participation, 6, 25, 64, 94
Pilot, pilot projects, 1–17, 23–56, 63–88, 93–100, 102, 106–108
Place, 25–27, 30, 32, 37, 38, 79, 103, 105
Policy, 3, 4, 6, 7, 17, 24, 27–30, 35–42, 45, 52, 66, 67, 69–74, 76, 82, 88, 93–108
Politicization, 53, 54
Politics, political, 3–8, 11, 14, 17, 24–26, 29, 30, 33–36, 38, 39, 43–46, 48, 52–56, 63, 65–68, 72, 75, 76, 79–88, 94, 99, 104, 106, 107

Practices, 2, 3, 5–7, 9, 11, 12, 14, 15, 17, 31–34, 42, 44–46, 48, 64–68, 70, 73, 74, 76–78, 80, 82, 85, 87, 94–98, 102–106
Production facilities, 15, 51
Prosumers, prosumption, 16, 27, 31, 70
Protective space, 9
Public engagement, 11, 12, 41, 75

R
Radical, 9, 10, 44, 50, 65, 74, 97, 100, 102
Regime, 9–12, 25, 30, 50, 64, 73, 102
Renewables, 3, 10, 13, 15–17, 25, 27, 36–38, 41, 50, 51, 68, 77, 78, 80, 96, 101
Replication, 47, 53
Research, 3, 4, 7, 10, 26, 27, 29, 30, 32, 36–38, 42, 44, 45, 66, 69–71, 73–75, 78, 86–88, 94, 96, 97, 99–102
Responsible Research and Innovation (RRI), 103

S
Scales, 3, 6, 15, 26, 38, 42, 48–53, 66, 67, 80
Science and technology studies (STS), 4, 7, 10–13, 17, 23, 24, 26, 30, 32, 46, 48, 52, 53, 64, 67, 71, 74, 75, 87, 96, 102, 105, 107, 108
Script, 71, 72
Shape, shaping, 3, 6, 10, 12, 13, 17, 24–26, 29–31, 35, 38–40, 43, 47, 54, 56, 67, 70, 72, 73, 82, 83, 87, 94, 108
Smart charging, 28, 80, 81, 83–86
Smart energy, 3, 6, 16, 24, 27–34, 36–38, 69–71, 79, 94–98, 100

Smart grid, 6, 16, 24, 25, 27–37, 68
Socio-technical, 2, 4, 5, 7–11, 13–15, 17, 25–27, 29–31, 33, 45, 48, 51–55, 63, 79, 93–108
Solar, solar power, 7, 15, 31, 49–51, 72, 77–79, 97
Storage, 2, 50, 51, 80
Strategic action field, 50, 77
Strategic energy technology (SET) plan, 95, 101

T
Target knowledge, 96, 97, 100
Technologies, 2–7, 9–17, 24–45, 47–56, 64–80, 84, 86–88, 93–108
Time-of use (TOU) tariffs, 16
Transformation, 14, 27, 36, 44, 46, 47, 51, 53, 54, 93–108
Transition, 2, 3, 5–17, 23, 25, 27, 30, 34, 36–40, 42–48, 50, 52–55, 64, 66, 68–70, 72–75, 77, 78, 80, 81, 83, 85, 86, 93–97, 99–101, 104–108
Translation, 46–48, 50, 53, 84
Transport, 2–4, 39–42, 44, 51, 55, 77, 80, 99, 101

U
Up-scaling, up-scale, 6, 12, 38, 45–49, 51–53
Urban, 35, 40, 41, 47, 49, 85, 86
Users, 2, 9, 10, 13, 31–34, 43, 47, 49, 55, 66, 68, 70–73, 75, 79, 96–98

V
Visions, 4, 28, 29, 38, 40, 41, 48–50, 53–55, 69–71, 85, 86, 95, 96, 105, 107, 108

The manufacturer's authorised representative in the EU is Springer Nature Customer Service Centre GmbH, Europaplatz 3, 69115 Heidelberg, Germany. If you have any concerns regarding our products, please contact ProductSafety@springernature.com

Printed and bound by CPI Group (UK) Ltd, Croydon, CR0 4YY

23/03/2026

02076443-0001